Regression Analysis
for Social Sciences

Regression Analysis for Social Sciences

Alexander von Eye
Department of Psychology
Michigan State University
East Lansing, Michigan

Christof Schuster
Institute for Social Research
University of Michigan
Ann Arbor, Michigan

Academic Press

San Diego   London   Boston   New York   Sydney   Tokyo   Toronto

This book is printed on acid-free paper. ∞

Academic Press
a division of Harcourt Brace & Company
525 B Street, Suite 1900, San Diego, California 92101-4495, USA
http://www.apnet.com

Academic Press Limited
24-28 Oval Road, London NW1 7DX, UK
http://www.hbuk.co.uk/ap/

Library of Congress Card Catalog Number: 98-84465

International Standard Book Number: 0-12-724955-9

PRINTED IN THE UNITED STATES OF AMERICA
98 99 00 01 02 03 BB 9 8 7 6 5 4 3 2 1

Contents

Preface

Regression analysis allows one to relate one or more criterion variables to one or more predictor variables. The term "to relate" is an empty formula that can be filled with an ever increasing number of concepts, assumptions, and procedures. Introductory textbooks of statistics for the social and behavioral sciences typically include only a description of classical simple and multiple regression where parameters are estimated using ordinary least-squares (OLS). OLS yields estimates that can have very desirable characteristics. Therefore it is deemed useful under many circumstances, as attested to by the many applications of this method. Indeed, OLS regression is one of the most frequently used methods of applied statistics.

Yet, there are many instances in which standard OLS regression is only suboptimal. In these instances, researchers may not wish to give up the goals of regressing one set of variables onto another and predicting values from the predictors. Instead, researchers may wish to apply more appropriate methods of regression, methods that are custom-tailored to their hypotheses, data, and assumptions.

A very large number of special methods for regression analysis are available. For instance, researchers can apply regression analysis to model dependent data or to depict curvilinear relationships. Researchers can apply various criteria when optimizing parameter estimates, or they can include categorical and continuous predictors simultaneously in a multiple regression equation. Most of these methods are easy to use, and many of these methods are already part of general purpose statistical software packages.

This book presents both the classical OLS approach to regression and modern approaches that increase the range of applications of regression considerably.

After the introduction in chapter one the second chapter explains the basics of regression analysis. It begins with a review of linear functions and explains parameter estimation and interpretation in some detail. Criteria for parameter estimation are introduced and the characteristics of OLS estimation are explained.

In the third chapter we extend the approach of simple, one-predictor regression to multiple regression. This chapter also presents methods of significance testing and a discussion of the multiple R^2, a measure used to estimate the portion of criterion variance accounted for by the regression model. The fourth chapter deals with simple and multiple regression using categorical predictors.

An important part of each regression analysis is outlier analysis. Outliers are extreme measures that either are far from the mean or exert undue influence on the slope of a regression equation. Both types of outliers are threats to the validity of results from regression analysis. Chapter 5 introduces readers to methods of outlier analysis. In a similar fashion, residual analysis is of importance. If residuals display systematic patterns, there is systematic variance that remains to be explained. Chapter 6 presents methods for residual analysis.

One example of systematic variability that standard regression using straight regression lines cannot detect is the presence of curvilinear relationships between predictors and criteria. Methods of polynomial regression, presented in Chapter 7, allow researchers to model virtually any shape of curve.

Multicollinearity can be a major problem in virtually any multiple regression analysis. The only exception is the use of perfectly orthogonal predictor variables. Multicollinearity problems result from dependence among predictors and can invalidate estimates of parameters. Chapter 8 presents methods for diagnosing multicollinearity and remedial measures.

Chapter 9 extends the curvilinear, polynomial approach to regression to multiple curvilinear regression. Another instance of nonlinear relationships is the presence of interactions between predictors. Multiplicative terms, which are routinely used to identify the existence of possible interactions, pose major problems for parameter interpretation. Specifically, multiplicative terms tend to create leverage points and multicollinearity problems and come with very uneven power which varies with the location of the interaction effect. Chapter 10 deals with interaction terms in regression.

When there is no obvious way to solve problems with outliers or the dependent variable is not normally distributed, methods of robust regression present a useful alternative to standard least-squares regression. Chapter 11 introduces readers to a sample of methods of robust regression.

Chapter 12 presents methods of symmetrical regression. These methods minimize a different goal function in the search for optimal regression parameters. Thus, this approach to regression allows researchers to elegantly deal with measurement error on both the predictor and the criterion side and with semantic problems inherent in standard, asymmetric regression. The chapter also presents a general model for ordinary least squares regression which subsumes a large array of symmetric and asymmetric approaches to regression and von Eye and Rovine's approach to robust symmetric regression.

In many instances researchers are interested in identifying an optimal subset of predictors. Chapter 13 presents and discusses methods for variable selection.

Of utmost importance in longitudinal research is the problem of correlated errors of repeatedly observed data. Chapter 14 introduces readers to regression methods for longitudinal data.

In many instances, the assumption of a monotonic regression line that is equally valid across the entire range of admissible or observed scores on the predictor cannot be justified. Therefore, researchers may wish to fit a regression line that, at some optimally chosen point on X, assumes another slope. The method of piecewise regression provides this option. It is presented in Chapter 15. The chapter includes methods for linear and nonlinear piecewise regression.

Chapter 16 presents an approach for regression when both the predictor and the criterion are dichotomous.

Chapter 17 illustrates application of regression analysis to a sample data set. Specifically, the chapter shows readers how to specify command files in SYSTAT that allow one to perform the analyses presented in the earlier chapters.

This volume closes with five appendices. The first presents elements of matrix algebra. This appendix is useful for the sections where least-squares solutions are derived. The same applies to Appendices B and C, which contain basics of differentiation and vector differentiation. Chapters 7 and 9, on polynomial regression, can be best understood with some elementary knowledge about polynomials. Appendix D summarizes this knowledge. The last appendix contains the data used in Chapter 14, on longitudinal data.

This book can be used for a course on regression analysis at the advanced undergraduate and the beginning graduate level in the social and behavioral sciences. As prerequisites, readers need no more than elementary statistics and algebra. Most of the techniques are explained step-by-step. For some of the mathematical tools, appendices are provided and can be consulted when needed. Appendices cover matrix algebra, the mechanics of differentiation, the mechanics of vector differentiation, and polynomials. Examples use data not only from the social and behavioral sciences, but also from biology. Thus, the book can also be of use for readers with biological and biometrical backgrounds.

The structure and amount of materials covered are such that this book can be used in a one-semester course. No additional reading is necessary. Sample command and result files for SYSTAT are included in the text. Many of the result files have been slightly edited by only including information of specific importance to the understanding of examples. This should help students and researchers analyze their own data.

We are indebted to a large number of people, machines, and structures for help and support during all phases of the production of this book. We can list only a few here. The first is the Internet. Most parts of this book were written while one of the authors held a position at the University of Jena, Germany. Without the Internet, we would still be mailing drafts back and forth. We thank the Dean of the Social Science College in Jena, Prof. Dr. Rainer K. Silbereisen, and the head of the Department of Psychology at Michigan State University, Prof. Dr. Gordon Wood for their continuing encouragement and support. We are grateful to Academic Press, notably Nikki Levy, for interest in this project. We are deeply impressed by Cheryl Uppling and her editors at Academic Press. We now know the intersection of herculean and sisyphian tasks.

We are most grateful to the creators of LaTeX. Without them, we would have missed the stimulating discussions between the two of us in which one of the authors claimed that all this could have been written in

WordPerfect in at least as pretty a format as in LaTeX. Luckily or not, the other author prevailed.

Most of all, our thanks are owed to Donata, Maxine, Valerie, and Julian. We also thank Elke. Without them, not one line of this book would have been written, and many other fun things would not happen. Thank you, for all you do!

December 1997

Alexander von Eye
Christof Schuster

Chapter 1

INTRODUCTION

Regression analysis is one of the most widely used statistical techniques. Today, regression analysis is applied in the social sciences, medical research, economics, agriculture, biology, meteorology, and many other areas of academic and applied science. Reasons for the outstanding role that regression analysis plays include that its concepts are easily understood, and it is implemented in virtually every all-purpose statistical computing package, and can therefore be readily applied to the data at hand. Moreover, regression analysis lies at the heart of a wide range of more recently developed statistical techniques such as the class of generalized linear models (McCullagh & Nelder, 1989; Dobson, 1990). Hence a sound understanding of regression analysis is fundamental to developing one's understanding of modern applied statistics.

Regression analysis is designed for situations where there is one continuously varying variable, for example, sales profit, yield in a field experiment, or IQ. This continuous variable is commonly denoted by Y and termed the *dependent variable*, that is, the variable that we would like to explain or predict. For this purpose, we use one or more other variables, usually denoted by $X_1, X_2, \ldots$, the *independent variables*, that are related to the variable of interest.

To simplify matters, we first consider the situation where we are only interested in a single independent variable. To exploit the information that the independent variable carries about the dependent variable, we try to find a mathematical function that is a good description of the as-

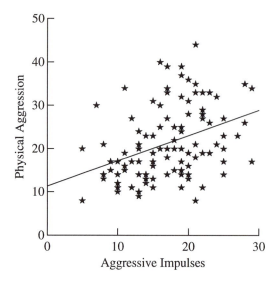

Figure 1.1: Scatterplot of aggressive impulses against incidences of physical aggression.

sumed relation. Of course, we do not expect the function to describe the dependent variable perfectly, as in statistics we always allow for randomness in the data, that is, some sort of variability, sometimes referred to as error, that on the one hand is too large to be neglected but, on the other hand, is only a nuisance inherent in the phenomenon under study.

To exemplify the ideas we present, in Figure 1.1, a scatterplot of data that was collected in a study by Finkelstein, von Eye, and Preece (1994). One goal of the study was to relate the self-reported number of aggressive impulses to the number of self-reported incidences of physical aggression in adolescents. The sample included $n = 106$ respondents, each providing the pair of values X, that is, Aggressive Impulses, and Y, that is, open Physical Aggression against Peers. In shorthand notation, $(X_i, Y_i), i = 1, \ldots, 106$.

While it might be reasonable to assume a relation between Aggressive Impulses and Physical Aggression against Peers, scientific practice involves demonstrating this assumed link between the two variables using data from experiments or observational studies. Regression analysis is one important tool for this task.

However, regression analysis is not only suited to suggesting decisions as to whether or not a relationship between two variables exists. Regression analysis goes beyond this decision making and provides a different type of precise statement. As we already mentioned above, regression analysis specifies a functional form for the relationship between the variables under study that allows one to estimate the degree of change in the dependent variable that goes hand in hand with changes in the independent variable. At the same time, regression analysis allows one to make statements about how certain one can be about the predicted change in Y that is associated with the observed change in X.

To see how the technique works we look at the data presented in the scatterplot of Figure 1.1. On purely intuitive grounds, simply by looking at the data, we can try to make statements similar to the ones that are addressed by regression analysis.

First of all, we can ask whether there is a relationship at all between the number of aggressive impulses and the number of incidences of physical aggression against peers. The scatterplot shows a very wide scatter of the points in the plot. This could be caused by imprecise measurement or a naturally high variability of responses concerning aggression. Nevertheless, there seems to be a slight trend in the data, confirming the obvious hypothesis that more aggressive impulses lead to more physical aggression. Since the scatter of the points is so wide, it is quite hard to make very elaborate statements about the supposed functional form of this relation. The assumption of a linear relation between the variables under study, indicated by the straight line, and a positive trend in the data seems, for the time being, sufficiently elaborate to characterize the characteristics of the data.

Every linear relationship can be written in the form $Y = \beta X + \alpha$. Therefore, specifying this linear relation is equivalent to finding reasonable estimates for β and α. Every straight line or, equivalently, every linear function is determined by two points in a plane through which the line passes. Therefore, we expect to obtain estimates of β and α if we can only find these two points in the plane. This could be done in the following way. We select a value on the scale of the independent variable, X, Aggressive Impulses in the example, and select all pairs of values that have a score on the independent variable that is close to this value. Now, a natural predictor for the value of the dependent variable, Y, Physical Aggression against Peers, that is representative for these observations is the mean of the dependent variable of these values. For example, when looking up in the scatterplot those points that have a value close to 10 on the Aggressive Impulse scale, the mean of the associated values on the physical aggression scale is near 15. Similarly, if we look at the points with a value close to 20 on the Aggressive Impulse scale, we find that the mean of the values of the associated Physical Aggression scale is located slightly above 20. So let us take 22 as our guess.

Now, we are ready to obtain estimates of β and α. It is a simple exercise to transform the coordinates of our hypothetical regression line, that is, (10, 15) and (20, 22), into estimates of β and α. One obtains as the estimate for β a value of 0.7 and as an estimate for α a value of 8. If we insert these values into the equation, $Y = \beta X + \alpha$, and set $X = 10$ we obtain for Y a value of 15, which is just the corresponding value of Y from which we started. This can be done for the second point, (20, 22), as well.

As we have already mentioned, the scatter of the points is very wide and if we use our estimates for β and α to predict physical aggression for, say, a value of 15 or 30 on the Aggressive Impulse scale, we do not expect it to be very accurate. It should be noted that this lack of accuracy is not caused by our admittedly very imprecise eyeballing method.

Of course, we do not advocate using this method in general. Perhaps the most obvious point that can be criticized about this procedure is that if another person is asked to specify a regression line from eyeballing, he or she will probably come to a slightly different set of estimates for α and β. Hence, the conclusion drawn from the line would be slightly different as well. So it is natural to ask whether there is a generally agreed-upon procedure for obtaining the parameters of the regression line, or simply the regression parameters. This is the case. We shall see that the regression parameters can be estimated optimally by the method of ordinary least squares given that some assumptions are met about the population the data were drawn from. This procedure will be formally introduced in the next chapters. If this method is applied to the data in Figure 1.1, the parameter estimates turn out to be 0.6 for β and 11 for α. When we compare these estimates to the ones above, we see that our intuitive method yields estimates that are not too different from the least squares estimates calculated by the computer.

Regardless of the assumed functional form, obtaining parameter estimates is one of the important steps in regression analysis. But as estimates are obtained from data that are to a certain extent random, these estimates are random as well. If we imagine a replication of the study, we would certainly not expect to obtain exactly the same parameter estimates again. They will differ more or less from the estimates of the first study. Therefore, a decision is needed as to whether the results are merely due to chance. In other words, we have to deal with the question of how likely it would be that we will not get the present positive trend in a replication study. It will be seen that the variability of parameter estimates depends not on a single factor, but on several factors. Therefore, it is much harder to find an intuitive reasonable guess of this variability then a guess of the point estimates for β and α.

With regression analysis we have a sound basis for examining the observed data. This topic is known as hypotheses testing for regression parameters or, equivalently, calculating confidence intervals for regression parameters. This topic will also be discussed in the following chapters.

Finally, having estimated parameters and tested hypotheses concerning these parameters, the whole procedure of linear regression rests on certain assumptions that have to be fulfilled at least approximately for the estimation and testing procedures to yield reliable results. It is therefore crucial that every regression analysis is supplemented by checking whether the inherent model assumptions hold. As an outline of these would not be reasonable without having formalized the ideas of estimation and hypothesis testing, we delay this discussion until later in this book. But it should be kept in mind that model checking is an essential part of every application of regression analysis.

After having described the main ideas of regression analysis, we would like to make one remark on the inconsistent use of the term "regression analysis." As can be seen from the example above, we regarded X, the independent variable, as a random variable that is simply observed. There are other situations where X is not a random variable but is determined prior to the study. This is the typical situation when data from a planned experiment are analyzed. In these situations the independent variable is usually under the experimenter's control and its values can therefore be set arbitrarily. Hence, X is fixed and not random. In both situations the analysis is usually referred to as regression analysis. One reason for this terminological oddity might be that the statistical analysis, as outlined in this book, does not vary with the definition of the randomness of X. With X fixed, or at least considered fixed, formulas are typically easier to derive because there is no need to consider the joint distribution of all the variables involved, only the univariate distribution of the dependent variable. Nevertheless it is important to keep these two cases apart as not all methods presented in this book can be applied to both situations. When X is fixed the model is usually referred to as the *linear model* (see, for instance, Searle, 1971; Graybill, 1976; Hocking, 1996). From this perspective, regression analysis and analysis of variance are just special cases of this linear model.

Chapter 2

SIMPLE LINEAR REGRESSION

Starting from linear functions, Section 2.1 explains simple regression in terms of *functional relationships between two variables*. Section 2.2 deals with parameter estimation and interpretation.

2.1   Linear Functions and Estimation

The following illustrations describe a functional relationship between two variables. Specifically, we focus on functions of the form

$$Y = \alpha + \beta X, \tag{2.1}$$

where Y is the criterion variable, β is some parameter, X is the predictor variable, and α is also some parameter. Parameters, variables, and their relationship can be explained using Figure 2.1.

The thick line in Figure 2.1 depicts the graph of the function $Y = \alpha + \beta X$, with $\beta = 2$, $\alpha = 1$, and X ranging between -1 and +3. Consider the following examples:

- for $X = 2$ we obtain $Y = 1 + 2 * 2 = 5$

- for $X = 3$ we obtain $Y = 1 + 2 * 3 = 7$

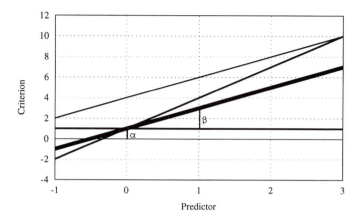

Figure 2.1: Functional linear relationship between variables Y and X.

- for $X = 0$ we obtain $Y = 1 + 2 * 0 = 1$

Using these examples and Figure 2.1 we find that

1. α is the *intercept parameter*, that is, the value assumed by Y when $x = 0$. For the example in the figure and the third numerical example we obtain $y = 1$ when $X = 0$, that is, $y = \alpha$; the level of the intercept is indicated as a horizontal line in Figure 2.1.

2. β is the *slope parameter*, that is, the increase in Y that is caused by an increase by one unit in X. For the example in the figure and the above numerical examples we find that when X increases by 1, for example, from $X = 2$ to $X = 3$, Y increases by 2, that is, $\beta = 2$.

3. Changes in α cause a parallel shift of the graph. Consider the upper thin line in Figure 2.1. This line is parallel to the thick line. The only difference between these two lines is that whereas the thinner line has an intercept of $\alpha = 4$, the thicker line has an intercept of $\alpha = 1$, or, the thicker line depicts the function $Y = 1 + 2X$, and the thinner line depicts the function $Y = 4 + 2X$. Positive changes in intercept move a curve upward. Negative changes in intercept move a curve downward.

4. Changes in β cause the angle that the line has with the X-axis to change. Consider the steepest line in Figure 2.1. This line has the

same intercept as the thick line. However, it differs from it in its slope parameter. Whereas the thick line increases by two units in Y for each unit increase in X. The thinner line increases by three units in Y for each unit in X, the thinner line depicts the function $Y = 1 + 3X$. Positive values of β cause the curve to increase as X increases. Negative values of β cause the curve to decrease as X increases.

Figure 2.1 depicts the relationship between X and Y so that X is the predictor and Y the criterion.

Consider a researcher that wishes to predict Cognitive Complexity from Reading Time. This researcher collects data in a random sample of subjects and then estimates a regression line. However, before estimating regression parameters, researchers must make a decision concerning the type or shape of regression line. For example a question is whether an empirical relationship can be meaningfully depicted using linear functions of the type that yield straight regression lines.[1] In many instances, this question is answered affirmatively.

When estimating regression lines, researchers immediately face the problem that, unlike in linear algebra, there is no one-to-one relationship between predictor values and criterion values. In many instances, one particular predictor value is responded to with more than one, different criterion values. Similarly, different predictor values can be responded to with the same criterion value. For instance, regardless of whether one drives on a highway in Michigan at 110 miles per hour or 120 miles per hour, the penalty is loss of one's driver license.

As a result of this less than perfect mapping, a straight line will not be able to account for all the variance in a data set. There will always be variance left unexplained. On the one hand, this may be considered a shortcoming of regression analysis. Would it not be nice if we could explain 100% of the variance of our criterion variable? On the other hand, explaining 100% of the variance is, maybe, not a goal worth pursuing. The reason for this is that not only in social sciences, but also in natural sciences, measurements are never perfect. Therefore, explaining 100% of

---

[1]The distinction between linear functions and linear regression lines is important. Chapters 7 and 9 of this book cover regression lines that are curvilinear. However, parameters for these lines are estimated using linear functions.

the variance either means that there is no error variance, which is very unlikely to be the case, or that the explanation confounded error variance with true variance.

While the claim of no error variance is hard to make plausible, the latter is hardly defensible. Therefore, we use for regression analysis a function that is still linear but differs from (2.1) in one important aspect: it contains a term for unexplained variance, also termed error or residual variance.

This portion of variance exists for two reasons. The first is that measures contain errors. This is almost always the case. The second reason is that the linear regression model does not allow one to explain all of the explainable variance. This is also almost always the case. However, whereas the first reason is a reason one has to live with, if one can, the second is curable, at least in part: one can try alternative regression models. But first we look at the *simple linear regression.*

The function one uses for simple linear regression, including the term for residual variance, is

$$y_i = \alpha + \beta x_i + \epsilon_i, \tag{2.2}$$

where

- y_i is the value that person i has on the criterion, Y.

- α is the intercept parameter. Later, we will show how the intercept can be determined so that it is the arithmetic mean of Y.

- β is the slope parameter. Note that neither α nor β have the person subscript, i. Parameters are typically estimated for the entire sample; only if individuals are grouped can two individuals have different parameter estimates.

- x_i is the value that person i has on the predictor, X.

- ϵ_i is that part of person i's value y_i that is not accounted for by the regression equation; this part is the residual. It is defined as

$$\epsilon_i = y_i - \mu_i, \tag{2.3}$$

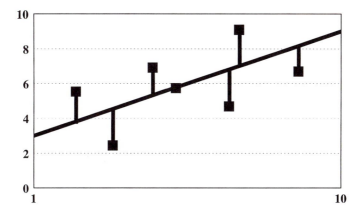

Figure 2.2: Illustration of residuals.

where μ_i is the value predicted using the regression equation (2.1), that is,

$$\mu_i = \alpha + \beta x_i. \tag{2.4}$$

Figure 2.2 displays a regression line and a number of data points around this line and the residuals. There is only one point that sits directly on the regression line. The other six data points vary in their distances from the regression line. Residuals are expressed in units of Y. Thus, the distance from the regression line, that is, the size of the residual, is depicted by the vertical lines. These lines are not the shortest connections to the regression lines. The shortest lines would be the lines that are orthogonal to the regression line and originate in the data points.

However, there is an explanation for the solution that expresses residuals in units of Y, that is, lines parallel to the Y-axis. The explanation is that differences between estimated values (regression line) and observed values (square data points) are best expressed in units of the criterion, Y (see Equation (2.2)). For example, weather forecasts typically include temperature forecasts for the coming days. It seems reasonable to express both the predicted temperature and the difference between measured and predicted temperatures in units of the temperature scale. Therefore, standard regression expresses residuals in units of Y. Chapter 12 of this book presents a solution for regression that expresses residuals in units of the orthogonal distance between the regression line and the data points.

Parameters α and β are not subscripted by individuals. This suggests that these parameters describe the entire population.

Basics about matrix algebra are given in Appendix A. This variant of algebra allows one to express algebraic expressions, in particular multiple equations, in an equivalent yet far more compact form. Readers not familiar with matrix algebra will benefit most from Section 2.2 and the following chapters if they make themselves familiar with it. Readers already familiar with matrix algebra may skip the excursus or use it as a refresher.

2.2   Parameter Estimation

This section is concerned with estimation and interpretation of parameters for regression equations. Using the methods introduced in Appendix B, we present a solution for parameter estimation. This solution is known as the *ordinary least squares solution.* Before presenting this solution, we discuss the method of least squares within the context of alternative criteria for parameter estimation.

2.2.1   Criteria for Parameter Estimation

A large number of criteria can be used to guide optimal parameter estimation. Many of these criteria seem plausible. The following paragraphs discuss five optimization criteria. Each of these criteria focuses on the size of the residuals, ϵ_i, where i indexes subjects.

The criteria apply to any form of regression function, that is, to linear, quadratic, or any other kinds of regression functions. One of the main characteristics of these criteria is that they are not statistical in nature. When devising statistical tests, one can exploit characteristics of the residuals that result from optimization according to one of certain criteria. However, the criteria themselves can be discussed in contexts other than statistics. For instance, the least squares criterion finds universal use. It was invented by Gauss when he solved the problem of constructing a road system that had to be optimal for military use.

All Residuals ϵ_i Must be Zero

More specifically, this criterion requires that

$$\epsilon_i = 0, \quad \text{for } i = 1, \ldots, n.$$

This first criterion may seem plausible. However, it can be met only under rare conditions. It implies that the relationship between X and Y is perfectly linear, and that the regression line goes exactly through each data point. As was discussed in Section 2.1, this is impractical for several reasons. One reason is that data are measured typically with error. Another reason is that, as soon as there is more than one case for values of X, researchers that employ this criterion must assume that the variation of Y values is zero, for each value of X. This assumption most probably prevents researchers from ever being able to estimate regression parameters for empirical data under this criterion.

Minimization of the Sum of Residuals

Let $\hat{y}_i$ be the estimated value for case i and y_i the observed value. Then, minimization of the sum of residuals can be expressed as follows:

$$Q = \sum_{i=1}^{n} (y_i - \hat{y}_i) = 0. \tag{2.5}$$

This criterion seems most reasonable. However, as it is stated, it can yield misleading results. One can obtain $Q = 0$ even if the regression line is the worst possible. This can happen when the differences $y_i - \hat{y}_i$ sum up to zero if their sums cancel each other out. This is illustrated in Figure 2.3

Figure 2.3 displays a string of data points. The first regression line suggests that, for these data points, the criterion given in (2.5) can be fulfilled. This line goes through each and every data point. As a result, we obtain for the sum of residuals $Q = 0$.

Unfortunately, the same applies for the second perpendicular line. The residuals for this line cancel each other out, because they lie at equal distances $y - \hat{y}$ from this regression line. As a result, the sum of residuals is $Q = 0$ too.

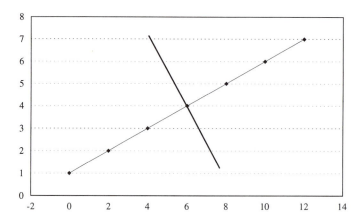

Figure 2.3: Two solutions for $Q = 0$.

Because of the problems that (1) solutions are not unique and (2) one has to devise criteria that allow one to discriminate between these two cases, this second criterion remains largely abandoned. The next criterion seems more practical.

Minimization of the Sum of the Absolute Residuals

To avoid the problems one can encounter when using (2.5), one can attempt minimization of the absolute residuals, $||y_i - \hat{y}_i||$, that is,

$$Q = \sum_{i=1}^{n} ||y_i - \hat{y}_i|| \rightarrow \min. \tag{2.6}$$

To illustrate Criterion (2.6), consider Figure 2.3 again. As a matter of course, the first regression line yields $Q = 0$. However, no other line yields this minimal value for Q. A large value for Q results for the perpendicular line.

While solving the biggest problem, that is, the problem of counter-intuitive regression slopes, Criterion (2.6) presents another problem. This problem concerns the uniqueness of solutions. Sole application of (2.6) cannot guarantee that solutions are unique. Solutions that are not unique, however, occur only rarely in analysis of empirical data. Therefore, Criterion (2.6) is occasionally applied and can be found in some statistical software packages (L1 regression in S+; Venables & Ripley, 1994).

Tshebysheff Minimization

Whereas the last three criteria focused on the sum of the residuals, the Tshebysheff Criterion focuses on individual residuals, ϵ_i. The criterion proposes that

$$\max ||\epsilon_i|| = \min.$$

In words, this criterion yields a curve so that the largest absolute residual is as small as possible. This is an intuitively sensible criterion. Yet, it is problematic for it is deterministic. It assumes that there is no error (or only negligible error) around measures. As a result, solutions from the Tshebysheff Criterion tend to severely suffer from outliers. One extreme value is enough to create bias, that is, dramatically change the slope of the regression line. Therefore, social science researchers rarely use this criterion. In contrast, the following criterion is the most frequently used.

Minimization of the squared Residuals: The Method of Ordinary Least Squares (OLS)

While solutions from Criterion (2.6) may be appropriate in many instances, lack of uniqueness of solutions minimizes enthusiasm for this criterion. Curiously, what was probably the idea that led to the development of OLS estimation[2] was the notion of distance between two points in n-space. Note that this distance is unique and it can also be easily calculated. It is well known that in a plane the distance between the two points p_1 and p_2 can be obtained by using the Pythagorean theorem through

$$d^2(p_1, p_2) = (x_{11} - x_{12})^2 + (x_{21} - x_{22})^2 = \sum_{i=1}^{2} (x_{i1} - x_{i2})^2,$$

---

[2]On a historical note, OLS estimation was created by the mathematician Gauss, the same person who, among other mathematical achievements, derived the formula for the normal distribution.

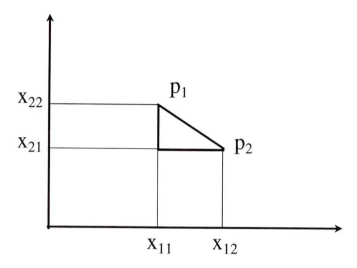

Figure 2.4: Distance between two points.

where p_1 and p_2 are two points in a plane with coordinates $p_1 = (x_{11}, x_{21})$ and $p_2 = (x_{12}, x_{22})$ and $d^2(p_1, p_2)$ denotes the squared distance between the two points. This is illustrated in Figure 2.4.

Now, suppose we have two points p_1 and p_2 in n-space, that is, $p_1 = (x_{11}, x_{21}, x_{31}, \ldots, x_{n1})$ and $p_2 = (x_{12}, x_{22}, x_{32}, \ldots, x_{n2})$. The squared distance is calculated as before with the sole exception that index i now ranges from 1 to n, that is,

$$d^2(p_1, p_2) = \sum_{i=1}^{n} (x_{i1} - x_{i2})^2.$$

The *least squares criterion* for parameter estimation in regression analysis has the following completely analogous form:

$$Q = \sum_{i=1}^{n} (y_i - \hat{y}_i)^2 \rightarrow \min. \tag{2.7}$$

In words, Criterion (2.7) states that we select the predicted values $\hat{y}$ such that Q, that is, the sum of the squared distances between the observations and the predictions, becomes a minimum. Of course, we will not select an arbitrary point in n-space for our prediction, because then we could chose $\hat{y} = y$ and the distance would always be zero. Only predictions that meet the model

$$\hat{y}_i = b_0 + b_1 x_i \tag{2.8}$$

are considered. Once the data are collected, the x_i values in (2.8) are known. Thus, the squared distance Q depends only on the values of b_0 and b_1. For every possible pair of b_0 and b_1 we could calculate Q, and the OLS principle states that we should select as reasonable estimates of the true but unknown regression coefficients β_0 and β_1 the values of b_0 and b_1 for which Q becomes a minimum.

Thus far we have given some reasons why OLS might be a sensible choice for estimating regression coefficients. In fact, when examining OLS estimators it turns out that they have a number of very desirable characteristics:

1. Least squares solutions are *unique*; that is, there is only one minimum

2. Least squares solutions are *unbiased*; that is, parameters estimated using least squares methods do not contain any systematic error

3. Least squares solutions are *consistent*; that is, when the sample size increases the solution converges toward the true population values

4. Least squares solutions are *efficient*; that is, their variance is finite and there is no solution with smaller variance among all unbiased linear estimators

Moreover, these estimators can be determined using closed forms. As a result, it is straightforward to computationally determine this minimum.

2.2.2 Least Squares Parameter Estimators

This chapter presents a solution to the least squares minimization problem, that is, it yields the estimates for regression parameters.

But before presenting the solution for estimation of regression parameters we take a closer and more formal look at the regression model and its characteristics.

The Regression Model: Form and Characteristics

As was stated in Section 2.1, the *population* model for simple regression can be formulated as

$$y_i = \beta_0 + \beta_1 x_i + \epsilon_i, \tag{2.9}$$

where y_i is subject i's value on the observed variable, β_0 and β_1 are the intercept and the slope parameters, respectively, x_i is a known constant, and ϵ_i is a random residual term.

This regression model has the following characteristics:

1. It is *simple*; that is, it involves only one predictor variable.

2. It is *linear in the parameters*; that is, parameters β_0 and β_1 have exponents equal to 1, do not appear as exponents, and are not multiplied or divided by any other parameter.

3. It is *linear in the predictor*; that is, the predictor also only appears in the first power. A model linear in its parameters and the predictor, that is, the independent variable, is also termed a *first-order linear model*.[3]

4. It is assumed that the random residual term, ϵ_i, has an expected mean of $E(\epsilon_i) = 0$ and variance $V(\epsilon_i) = \sigma^2$. In words, the residuals are assumed to have a mean of zero and a constant variance.

5. The residuals for two different observations i and j are assumed to be independent. Therefore, their covariance is zero, that is, $C(\epsilon_i, \epsilon_j) = 0$.

6. Variable X is a constant.[4]

7. The regression model contains a systematic component, $\beta_0 + \beta_1 x_i$, and a random term, ϵ_i. Therefore, the observed y_i is a random variable.

---

[3]It should be noted that, in curvilinear regression, predictor variables can appear in nonlinear form. However, as long as the parameters appear only in the first power, the regression model can be termed linear.

[4]Remember the distinction between fixed and random predictors in the Introduction.

8. The observed value, y_i, differs from the value expected from the regression equation by an amount given by ϵ_i.

9. Because the residual terms, ϵ_i, are assumed to have constant variance, the observed values, y_i, have the same constant variance:

$$V(y_i) = \sigma^2.$$

This implies that the variance of Y does not change with X.

10. Because the residual covariance $C(\epsilon_1, \epsilon_j) = 0$, for $i \neq j$, the responses are uncorrelated also:

$$C(y_1, y_j) = 0, \text{ for } i \neq j.$$

11. Y must be real-valued. The regression equation given in (2.9) can yield estimates that are positive, negative, or zero. In addition, (2.9) can yield predicted values at virtually any level of resolution. If the criterion cannot assume these values, the linear regression model in tandem with OLS estimators may not be the appropriate model. There are no such constraints placed on X. X can be even categorical (see Chapter 4 of this volume).

In order to obtain the estimates of the regression coefficients the easy way, we need some calculus that is not covered in the excursus on matrix algebra. The excursus in Appendix B reviews this calculus (for more details see, e.g., Klambauer, 1986). Readers well trained in calculus may wish to skip the excursus.

It should be noted that, for valid estimation of regression parameters, neither X nor Y is required to be normally distributed. In contrast, the significance tests and the construction of confidence intervals developed later require that Y be normally distributed.

Estimating Regression Slope and Intercept Parameters

This section presents the standard OLS solution for the slope and intercept regression parameters. Recall from the above section that the function to be minimized is

$$Q = \sum_{i=1}^{n}(y_i - \hat{y}_i)^2 \to \min. \tag{2.10}$$

The estimated Y value for case i is

$$\hat{y}_i = b_0 + b_1 x_i. \tag{2.11}$$

Inserting (2.11) into (2.10) yields

$$Q(b_0, b_1) = \sum_{i=1}^{n}(y_i - b_0 - b_1 x_i)^2.$$

Note that we consider $Q(b_0, b_1)$ as a function of b_0 and b_1 and not as a function of y_i and x_i. The reason for this is that once the data are collected they are known to us. What we are looking for is a good guess for b_0 and b_1; therefore, we treat b_0 and b_1 as variables. We are now taking the two partial derivatives of $Q(b_0, b_1)$. After this, all that remains to be done is to find that point (b_0, b_1) where these partial derivatives both become zero. This is accomplished by setting them equal to zero, yielding two equations in two unknowns. These equations are then solved for b_0 and b_1. The solutions obtained are the OLS estimates of β_0 and β_1.

First we take the partial derivative of $Q(b_0, b_1)$ with respect to b_0. Interchanging summation and differentiation and then applying the chain rule,

$$\begin{aligned}
\frac{\partial}{\partial b_0}Q(b_0, b_1) &= \frac{\partial}{\partial b_0}\sum_{i=1}^{n}(y_i - b_0 - b_1 x_i)^2 \\
&= \sum_{i=1}^{n}\frac{\partial}{\partial b_0}(y_i - b_0 - b_1 x_i)^2 \\
&= \sum_{i=1}^{n}-2(y_i - b_0 - b_1 x_i) \\
&= -2\sum_{i=1}^{n}(y_i - b_0 - b_1 x_i).
\end{aligned}$$

The last equation is set to zero and can then be further simplified by dropping the factor -2 in front of the summation sign and resolving the expression enclosed by parentheses,

$$
\begin{aligned}
0 &= \sum_{i=1}^{n}(y_i - b_0 - b_1 x_i) \\
&= \sum_{i=1}^{n} y_i - n b_0 - b_1 \sum_{i=1}^{n} x_i \\
&= n\bar{y} - n b_0 - n b_1 \bar{x}.
\end{aligned}
$$

Note that in this derivation we have used the fact that $\sum_{i=1}^{n} x_i = n\bar{x}$. This applies accordingly to the y_i. Finally solving for b_0 yields

$$
b_0 = \bar{y} - b_1 \bar{x}. \tag{2.12}
$$

It can be seen that we need b_1 to obtain the estimate of b_0. Now we take the partial derivative of $Q(b_0, b_1)$ with respect to b_1, that is,

$$
\begin{aligned}
\frac{\partial}{\partial b_1} Q(b_0, b_1) &= \frac{\partial}{\partial b_1} \sum_{i=1}^{n}(y_i - b_0 - b_1 x_i)^2 \\
&= \sum_{i=1}^{n} \frac{\partial}{\partial b_1}(y_i - b_0 - b_1 x_i)^2 \\
&= \sum_{i=1}^{n} -2 x_i (y_i - b_0 - b_1 x_i) \\
&= -2 \sum_{i=1}^{n} x_i (y_i - b_0 - b_1 x_i).
\end{aligned}
$$

Again, setting this equation to zero and simplifying, we obtain

$$
\begin{aligned}
0 &= \sum_{i=1}^{n} x_i (y_i - b_0 - b_1 x_i) \\
&= \sum_{i=1}^{n} x_i y_i - b_0 \sum_{i=1}^{n} x_i - b_1 \sum_{i=1}^{n} x_i^2 \\
&= n\bar{y} - n b_0 - n b_1 \bar{x}.
\end{aligned}
$$

Substituting the expression for b_0 from (2.12) yields

$$
\begin{aligned}
0 &= \sum_{i=1}^{n} x_1 y_i - (\bar{y} - b_1 \bar{x}) \sum_{i=1}^{n} x_i - b_1 \sum_{i=1}^{n} x_i^2 \\
&= \sum_{i=1}^{n} x_i y_i - n\bar{x}\bar{y} + b_1 \left[\frac{1}{n} \left(\sum_{i=1}^{n} x_i \right)^2 - \sum_{i=1}^{n} x_i^2 \right].
\end{aligned}
$$

Finally, solving for b_1,

$$
b_1 = \frac{\sum_{i=1}^{n} x_i y_i - n\bar{x}\bar{y}}{\sum_{i=1}^{n} x_i^2 - \frac{1}{n} \left(\sum_{i=1}^{n} x_i \right)^2}.
$$

Regression parameters are thus obtained by first calculating b_1 and then using this estimate in the formula for calculating b_0. It should be noted that from a purely mathematical viewpoint we would have to go a step further in proofing that the parameter estimates just obtained indeed give the minimum of the function surface $Q(b_0, b_1)$ in three-dimensional space and not a maximum or a saddle-point. But we can assure the reader that this solution indeed describes the minimum. With further algebra the slope parameter, b_1, can be put in the following form:

$$
b_1 = \frac{\sum_{i=1}^{n} (x_i - \bar{x})(y_i - \bar{y})}{\sum_{i=1}^{n} (x_i - \bar{x})^2}. \tag{2.13}
$$

The numerator of (2.13) contains the cross product of the centered[5] variables X and Y, defined as the inner product of the vectors $(\mathbf{x} - \bar{\mathbf{x}})$ and $(\mathbf{y} - \bar{\mathbf{y}})$, that is, $(\mathbf{x} - \bar{\mathbf{x}})'(\mathbf{y} - \bar{\mathbf{y}})$ estimation.

Data Example

The following numerical example uses data published by Meumann (1912). The data describe results from a memory experiment. A class of children in elementary school (sample size not reported) learned a sample of 80 nonsense syllables until each student was able to reproduce 100% of the

---

[5] Centering a variable is defined as subtracting the variable's arithmetic mean from each score.

syllables. Five minutes after everybody had reached this criterion, students were asked to recall the syllables again. This was repeated a total of eight times at varying time intervals. After some appropriate linearizing transformation trials were equidistant.

Because the experimenter determined when to recall syllables, the independent variable is a fixed predictor. At each time the average number of recalled syllables was calculated. These are the data that we use for illustration.

We regress the dependent variable Y, number of nonsense syllables recalled, onto the independent variable X, trial without intermittent repetition.[6] Table 2.1 presents the raw data and the centered variable values.

Table 2.1 shows the raw data in the left-hand panel. The first column contains the Trial numbers, X. The second column contains the average recall rates, Y. The mean value for the trial number is $\bar{x} = 4.5$, while the mean for the recall rate is $\bar{y} = 52.32$.

The next four columns illustrate the numerical steps one needs to perform for calculation of parameter estimates. The model we investigate for the present example can be cast as follows:

$$\text{Recall} = \beta_0 + \beta_1 * \text{Trial} + \epsilon. \tag{2.14}$$

The results displayed in the first four columns of the right panel are needed for estimating the slope parameter, b_1, using Equation (2.13). We use this formula for the following calculations. The first of these columns contains the centered trial variable. We calculate these values by subtracting from each trial number the mean of the trials, that is, $\bar{x} = 4.5$. The next column contains the square of the centered trial numbers. We need the sum of these values for the denominator of Formula (2.13). This sum appears at the bottom of this column.

The fifth column contains the centered recall rates. The summands of the cross product of the centered predictor, *Trials*, and the centered

---

[6] Considering the well-known, nonlinear shape of decay, that is, forgetting curves, one may wonder whether the distances between recall trials were equal. Without having access to the original data, we assume that the time distance between recall trials increased systematically with the number of trials. If this assumption is correct, one cannot validly regress recall rates onto time, unless one possesses information about the time passed between the trials.

Table 2.1: *Estimation of Slope Parameter for Data from Forgetting Experiment*

Variables		Calculations Needed for Estimation of b_1				
x_i	y_i	$x_i - \bar{x}$	$(x_i - \bar{x})^2$	$y_i - \bar{y}$	$(y_i - \bar{y})^2$	$z_i^x z_i^{y\,a}$
1	78.00	−3.5	12.25	25.68	659.46	−89.88
2	70.88	−2.5	6.25	18.56	344.47	−46.40
3	56.56	−1.5	2.25	4.24	17.98	−6.36
4	37.92	−0.5	0.25	−14.40	207.36	7.20
5	54.24	0.5	0.25	1.92	3.69	0.96
6	48.72	1.5	2.25	−3.60	12.96	−5.40
7	39.44	2.5	6.25	−12.88	165.89	−32.20
8	32.80	3.5	12.25	−19.52	381.03	−68.32
Sums			42.00		1792.84	−240.40

[a]The variables z_i^x and z_i^y denote the centered x and y values, respectively. For instance, $z_i^x = (x_i - \bar{x})$.

criterion, *Recall*, appear in the last column of Table 2.1. The sum of the cross products appears at the bottom of this column. It is needed for the numerator of Formula (2.13).

Using the calculated values from Table 2.1 in Formula (2.13) we obtain

$$b_1 = b_1 = \frac{\sum_{i=1}^{n}(x_i - \bar{x})(y_i - \bar{y})}{\sum_{i=1}^{n}(x_i - \bar{x})^2} = \frac{-240.4}{42} = -5.72.$$

For the intercept parameter we obtain

$$b_0 = \bar{y} - b_1\bar{x} = 52.3 - (-5.72) * 4.5 = 78.04.$$

Figure 2.5 presents the Trial x Recalled Syllables plot. It also contains the regression line with the parameters just calculated.

Figure 2.5 suggests that, with Trial 4 being the only exception, the linear regression line provides a good rendering of the decay of the learned material.

In order to quantify how well a regression line depicts a relationship most researchers regress using methods that include statistical significance tests and measures of variance accounted for. The former are presented in Section 2.5. One example of the latter is presented in the following

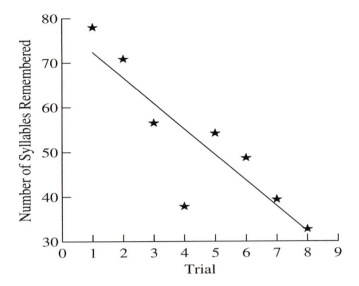

Figure 2.5: Scatterplot showing the dependence of the recall performance on trial.

section.

2.2.3 The Goodness of Fit of the Regression Model

One measure of how well a statistical model explains the observed data is the *coefficient of determination*, that is, the square of the Pearson correlation coefficient, r, between y and $\hat{y}$. This measure describes the percentage of the total variance that can be explained from the predictor-criterion covariance. In its general form, the correlation coefficient is given by

$$r = \frac{\sum_i (x_i - \bar{x})(y_i - \bar{y})}{\sqrt{\sum_i (x_i - \bar{x})^2 \sum_i (y_i - \bar{y})^2}}.$$

If we replace x with $\hat{y}$ in the above formula we obtain the multiple correlation R. The reasons for making a distinction between r and R are that

- r is a measure of association between two random variables whereas

R is a measure between a random variable y and its prediction $\hat{y}$ from a regression model.

- r lies in the interval $-1 \leq r \leq 1$ while the multiple correlation R cannot be negative; that is, it lies in the interval $0 \leq R \leq 1$.

- R is always well defined, regardless of whether the independent variable is assumed to be random or fixed. In contrast, calculating the correlation between a random variable, Y, and a fixed predictor variable, X, that is, a variable that is not considered random, makes no sense.

The square of the multiple correlation, R^2, is the coefficient of determination. If the predictor is a random variable the slope parameter can be expressed as

$$b_1 = \frac{\sum_i (y_i - \bar{y})^2}{\sum_i (x_i - \bar{x})^2} \, r = \frac{s_y^2}{s_x^2} \, r,$$

where s_x^2 and s_y^2 are the variances of the predictor and the criterion, respectively. If the predictor, X, is a random variable it makes sense to calculate the correlation between X and Y. Then, R^2 is identical to r^2 as $\hat{Y}$ is just a linear transformation of X, and correlations are invariant against linear transformations.

Alternatively, R^2 can be expressed as the ratio of variance explained by regression, SSR, and the total variance, $SSTO$, or

$$R^2 = \frac{SSR}{SSTO} = 1 - \frac{SSE}{SSTO}, \tag{2.15}$$

where SSE is the variance of the residuals. The right-hand term in (2.15) suggests that R^2 is a measure of proportionate reduction in error. In other words, R^2 is a measure of the proportionate reduction in the variability of Y that can be accomplished by using predictor X.

2.3   Interpreting Regression Parameters

This section presents aids for interpretation of the regression parameters b_0 and b_1. We begin with b_0.

2.3.1 Interpreting the Intercept Parameter

Formula (2.12) suggests that b_0 is composed of two summands. The first summand is the arithmetic mean of Y, $\bar{y}$. The second is the product of the slope parameter, b_1, and the arithmetic mean of X, $\bar{x}$. Now suppose that X is centered, that is, the arithmetic mean of X has been subtracted from each value x_i. Centering transforms the array of x values as follows:

$$\begin{pmatrix} x_1 \\ x_2 \\ \vdots \\ x_n \end{pmatrix} \rightarrow \begin{pmatrix} x_1 - \bar{x} \\ x_2 - \bar{x} \\ \vdots \\ x_n - \bar{x} \end{pmatrix}.$$

Centering constitutes a parallel shift in the data. Parallel shifts are linear transformations. Regression slopes are invariant against parallel shifts. The effect of centering is that the mean of an array of data is shifted to be zero. Centering X has the effect that the second summand in (2.12) is zero. Therefore, whenever X is centered, the intercept parameter equals the arithmetic mean of Y. When X is not centered, the intercept parameter equals the difference between the arithmetic mean of Y and the product of the slope parameter and the arithmetic mean of X.

Consider the example given in Table 2.1. The criterion in this example, Recall, had an arithmetic mean of $\bar{y} = 418.56/8 = 52.3$. The intercept parameter was $b_0 = 78.04$. We conclude that X had not been centered for this example. The difference between 52.3 and 78.04 must then be $b_1\bar{x} = -5.72 * 4.5 = -25.74$. Adding this value to 52.3 yields 78.04. Readers are invited to recalculate b_0 using the centered values for *Trial* in the third column of Table 2.1.

The interpretation of intercept parameters across samples is possible, in particular when X has been centered. It has the same implications as mean comparisons across samples.

2.3.2 Interpreting the Slope Parameter

Slope parameters indicate how steep the increase or decrease of the regression line is. However, there is a simpler interpretation. The slope parameter in simple regression indicates the number of steps one moves on Y when taking one step on X.

Consider again the example in Table 2.1. For this example we calculated the following regression equation:

$$\text{Recall} = 78.04 - 5.72 * \text{Trial} + \text{Residual}. \tag{2.16}$$

Inserting a value from the interval between Trial = 1 and Trial = 8 yields estimates of Recall that are located on the regression line in Figure 2.5. For example, for Trial = 3 we calculate an average estimated Recall = 60.91. Taking one step on X, from Trial = 3 to Trial = 4, yields the estimate for Recall of 55.18. The difference between these two estimates, $60.91 - 55.18 = -5.72$, is exactly the estimated regression slope.

2.4   Interpolation and Extrapolation

Interpolation is defined as estimating Y values for X values that are within the interval between the maximum and the minimum X values used for estimating regression parameters. Specifically, interpolation refers to X values not realized when estimating regression parameters. In many instances, interpolation is a most useful and meaningful enterprise. Assuming linear relationships, one can use interpolation to estimate the effects of some intermediate amount of a drug or some intermediate amount of a fertilizer.

In these examples, the predictor is defined at the ratio scale level, and assuming fractions of predictor values is reasonable. The same applies to predictors defined at the interval level. In other instances, however, when predictors are defined at the ordinal or nominal scale levels, interpolation does not make much sense.

Consider the example in Table 2.1 as illustrated in Figure 2.5. This example involves a predictor at the ratio scale level. For instance, Trial 3.3 indicates that a third of the time between the 3rd and the 4th trial has passed. Therefore, interpolation using the trial scale is meaningful. Thus, if the trial variable is closely related to some underlying variable such as time, interpolation can make sense.

Extrapolation is defined as the estimation of Y values for X values beyond the range of X values used when estimating regression parameters. As far as scale levels are concerned, the same caveats apply to extrapolation as to interpolation. However, there is one more caveat.

Extrapolation, when not sensibly performed, can yield results that are implausible, or worse. Predictor values that are conceivable can yield criterion values that simply cannot exist. The predicted values may be physically impossible or conceptually meaningless. Consider the regression equation given in (2.16). Inserting predictor values using the natural numbers between 1 and 8 yields reasonable estimates for Recall. Inserting natural numbers beyond these boundaries can yield implausible results.

For example, consider a hypothetical Trial 10. From our regression equation we predict a recall rate of 20.84. This is conceptually meaningful and possible. Consider, however, hypothetical Trial 15. Inserting into our regression equation results in an estimated recall rate of -7.78. This value is conceptually meaningless. One cannot forget more than what one had in memory.

Thus, while extrapolating may use conceptually meaningful predictor values, the resulting estimates must be inspected and evaluated as to their meaningfulness. The same applies to predictor values. In the present example, there can be no trial number -5. We can insert this number into the equation and calculate the meaningless estimated recall rate of 106.70 out of 80 items on the list. However, regardless of how conceptually meaningful the estimate will be, when predictor values are impossible, results cannot be interpreted.

2.5 Testing Regression Hypotheses

The two most important methods for evaluating how well a regression line describes the predictive relationship between predictor X and criterion Y focus on the proportion of variance accounted for, R^2, and on statistical significance. R^2 was explained in Section 2.2.3. Recall that R^2 describes how valuable the regression model is for predicting Y from knowledge of X. If R^2 is near 1, the observations in the XY scatterplot lie very close to the regression line. If R^2 is near 0, the scatter around the regression line is very wide.

The present section introduces readers to statistical significance testing. Three types of hypotheses are covered. The first concerns a single regression slope coefficient,[7] b_1. The hypothesis to be tested is whether β_1 is different than either zero (Section 2.5.1) or equal to some constant (Section 2.5.3). The second hypothesis concerns the comparison of two slope regression coefficients from independent samples (Section 2.5.4). The third hypothesis asks whether the intercept coefficient is different than a constant (Section 2.5.6). Finally, the construction of confidence intervals for regression coefficients is covered in Section 2.5.7. Chapter 3, concerned with multiple regression, covers additional tests.

However, before presenting hypotheses and significance tests in detail, we list conditions that must be fulfilled for significance tests to be valid and meaningful. These conditions include:

1. The residuals must be normally distributed.

2. The criterion variable must be measured at the interval level or at the ratio scale level.

3. Researchers must have specified a population.

4. The sample must be representative of this population.

5. Members of the sample must be independent of each other.

6. The sample must be big enough to give the alternative hypothesis a chance to prevail (for power analyses, see Cohen, 1988).

7. The sample must be small enough to give the null hypothesis a chance to prevail.

8. The matrix $\mathbf{X}$ is assumed to be of full column rank; that is, the columns of the matrix $\mathbf{X}$ must be linearly independent of each other. This is typically the case if the columns of matrix $\mathbf{X}$ represent numerical variables. For categorical variables full column rank can always be achieved by suitable reparameterization. Therefore, $\mathbf{X}$ can typically be assumed to be of full column rank.

---

[7]The term regression coefficient is used as shorthand for estimated regression parameter.

2.5.1 Null Hypothesis: Slope Parameter Equals Zero

The most frequently asked question concerning statistical significance of the regression slope coefficient is whether the regression coefficient is significant. In more technical terms, this question concerns the hypothesis that the slope coefficient describes a sample from a population with a slope parameter different from zero. In null hypothesis form one formulates

$$H_0 : \beta_1 = 0.$$

There are three possible alternative hypotheses to this null hypothesis. First there is the two-sided alternative hypothesis

$$H_1 : \beta_1 \neq 0.$$

Using this hypothesis researchers do not specify whether they expect a regression slope to go upward (positive parameter) or downward (negative parameter). The null hypothesis can be rejected in either case, if the coefficient is big enough. In contrast, when researchers expect a slope parameter to be positive, they test the following, one-sided alternative hypothesis

$$H_1 : \beta_1 > 0.$$

Alternatively, researchers specify the following one-sided alternative hypothesis when they expect the regression slope to be negative:

$$H_1 : \beta_1 < 0.$$

In the memory experiment analyzed in Section 2.2.2, the obvious hypothesis to be tested is $H_1 : \beta_1 < 0$, because we expect recall rates to decrease over time.

Statistical tests of the two-sided alternative hypothesis are sensitive to large slope coefficients, regardless of sign. In contrast, tests of the one-sided alternative hypotheses are sensitive only to slope coefficients with the right sign. Whenever the null hypothesis can be rejected, researchers conclude that the sample was drawn from a population with slope parameter $\beta_1 > 0$, $\beta_1 < 0$, or $\beta_1 \neq 0$, depending on what null hypothesis was tested. This conclusion is based on a probability statement.

This statement involves two components, the significance level, α, and the population. The significance level is typically set to $\alpha = 0.05$ or $\alpha = 0.01$. If the null hypothesis can be rejected, the probability is less than α that the sample was drawn from a population with parameter $\beta_1 = 0$. Obviously, this statement makes sense only if the population was well defined and the sample is representative of this population (see Condition 3, above). The null hypothesis prevails only if there is no (statistically significant) way to predict the criterion from the predictor.

In the following we present an F test that allows one to test the above hypotheses. In Chapter 3, when we cover multiple regression, we present a more general version of this test. The F test for simple regression can be given as follows:

$$F = \frac{r^2(n-2)}{1-r^2}, \quad \text{for } r^2 < 1. \tag{2.17}$$

The F test given in (2.17) relates two fractions to each other. The fraction in the numerator involves the portion of variance that is accounted for by the linear regression of Y on X, r^2. This portion, given by the coefficient of determination, is weighted by the numerator's degrees of freedom, $df_1 = 2 - 1 = 1$, that is, the number of predictors (including the constant) minus one. The fraction in the denominator involves the portion of variance that remains unaccounted for by the linear regression of Y on X, $1 - r^2$. This portion is weighted by the denominator's degrees of freedom, $df_2 = n - 2$, that is, the sample size minus the number of predictors (including the constant). In brief, this F ratio relates the weighted variance accounted for to the weighted unexplained variance.

Formula (2.17) also contains a constraint that is important when employing tests of statistical significance. This constraint requires that less than 100% of the variance be accounted for. If 100% of the variance is accounted for, and the authors of this book have yet to see a case where empirical data can be explained to 100%, there is no need for statistical testing. If the residual variance is zero, there is nothing to test against. In this case, one may consider the relationship between the predictor and the criterion deterministic and may skip the significance test.

When deciding whether or not to reject the null hypothesis, one proceeds as follows: If $F \geq F_{\alpha, df1, df2}$ then reject H_0. This F test is two-sided.

Many statistical software packages, for example, SYSTAT and SAS, use a t test instead of this F test. These two tests are equivalent with no differences in power. The t test for the hypothesis that $\beta_1 = 0$ has the form:

$$t = \frac{b_1}{s(b_1)}, \tag{2.18}$$

where $s(b_1)$ denotes the estimated standard deviation of b_1. The estimated variance of b_1 is

$$s^2(b_1) = \frac{MSE}{\sum_{i=1}^n (x_i - \bar{x})^2} = \frac{\frac{1}{n-2} \sum_{i=1}^n e_i^2}{\sum_{i=1}^n (x_i - \bar{x})^2}, \tag{2.19}$$

where the mean squared error, MSE, is the sum of the squared residuals divided by its degrees of freedom, $n - 2$. The degrees of freedom of this t test appear in the denominator of the MSE.

2.5.2 Data Example

The following numerical example uses data from the Vienna Longitudinal Study on cognitive and academic development of children (Spiel, 1998). The example uses the variables Performance in Mathematics in Fourth Grade, *M4* and Fluid Intelligence, *FI*. Using data from a sample of $n = 93$ children, representative of elementary school children in Vienna, Austria, we regress *M4* on *FI*. Figure 2.6 displays the raw data and the linear regression line.

The regression function for these data is

$$M4 = 2.815 + 0.110 * FI + \text{Residual}.$$

The correlation between *M4* and *FI* is $r = 0.568$, and $r^2 = 0.323$. We now ask whether the regression of *M4* on *FI* has a slope parameter that is statistically significantly different than zero. We calculate F by inserting into (2.17),

$$F = \frac{0.323 * 91}{1 - 0.323} = 43.417.$$

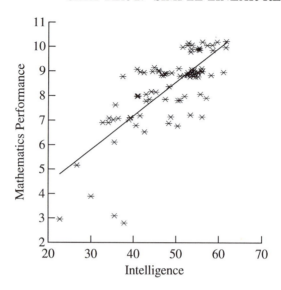

Figure 2.6: Scatterplot showing the dependence of the performance in Mathematics on Intelligence.

The critical test statistic for a two-sided test is $F_{0.05,1,91} = 3.94$. The empirical $F = 43.417$ is greater than this value. We therefore conclude that the null hypothesis of a zero slope parameter can be rejected, and favor the alternative hypothesis that the slope parameter is different than zero.

The t value for the same hypothesis is $t = 6.589$. Squaring yields $t^2 = 43.417$, that is, the same value as was calculated for F. This illustrates the equivalence of t and F.

2.5.3   Null Hypothesis: Slope Parameter Constant

This section presents a t test for the null hypothesis that an empirical slope coefficient is equal to some *a priori* determined value, k. The t statistic is of the same form as that of the statistic for the hypothesis that $\beta_1 = 0$ and can be described as

$$t = \frac{b_1 - k}{s(b_1)}, \qquad (2.20)$$

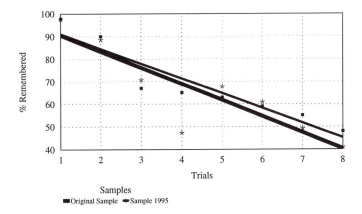

Figure 2.7: Comparing two samples in the forgetting experiment.

where $s(b_1)$ is defined as in Equation (2.19).

The following data example compares results from the Meumann (1912) forgetting experiment (Section 2.2.2) with a hypothetical sample from 1995. Participants in the replication experiment learned the same syllables until they were able to recall 100%. Forgetting was tested using the same number of trials in the same time intervals.

Figure 2.7 displays the original data together with the data from the replication experiment. The data for the new sample appear also in Table 2.2. The data for the original sample appear in Table 2.1.

The figure is arranged so that forgetting is regressed onto trials as in Figure 2.7. The figure suggests that forgetting is less brisk in the replication sample than in the original sample.

We test the hypothesis that the slope parameter for the regression of Recall on Trial in the 1995 sample is $\beta_1 = -10$. Inserting the values from the table into (2.20) yields

$$t = \frac{-6.543 + 10}{\sqrt{\frac{2056.96}{6}}}\sqrt{42} = 1.21.$$

The tail probability for this value is, for $df = 6, p = 0.2718$ (two-sided test). This value is greater than $\alpha = 0.05$. Thus, we cannot reject the null hypothesis that the slope coefficient in the population is equal to

Table 2.2: *Estimation of Slope Parameter for Data from Replication of Forgetting Experiment: Comparison of Two Samples*

Variables		Calculations Needed for Estimation of b_1				
x_i	y_i	$x_i - \bar{x}$	$(x_i - \bar{x})^2$	$y_i - \bar{y}$	$(y_i - \bar{y})^2$	$z_i^x z_i^{y\,a}$
1	97.8	-3.5	12.25	29.7	882.09	-103.95
2	90.0	-2.5	6.25	21.9	479.61	-54.75
3	67.0	-1.5	2.25	-1.1	1.21	1.65
4	65.0	-0.5	0.25	-3.1	9.61	1.55
5	63.0	0.5	0.25	-5.1	26.01	-2.55
6	59.0	1.5	2.25	-9.1	82.81	-13.65
7	55.0	2.5	6.25	-13.1	171.61	-32.75
8	48.0	3.5	12.25	-20.1	404.01	-70.35
Sums			42.00		2056.96	-274.80

  [a]The variables z_i^x and z_i^y denote the centered x and y values, respectively. For instance, $z_i^x = (x_i - \bar{x})$.

$\beta_1 = -10$.

2.5.4 Null Hypothesis 2: Two Slope Parameters are Equal

Many researchers consider it desirable to replicate studies using independent samples. Results are deemed more trustworthy when replication studies exist that confirm results of earlier studies. Comparing regression coefficients can be an important part of comparing studies.

Nobody expects that two estimated regression coefficients, b_1 and b_2, from independent samples are numerically exactly the same. Rather, one has two expectations:

1. The significance status (significant versus insignificant) of the two coefficients, b_1 and b_2, is the same.

2. The numerical difference between the two estimates is "small"; that is, the two coefficients are not significantly different from each other.

Whereas the first expectation can be tested using individual tests of b_1 and b_2, the second involves a comparison of slope coefficients. This comparison can be performed using the following t test. Let b_1 be the slope coefficient estimated for the first sample, and b_2 the slope coefficient for the second, independent sample. Then, the t test for the null hypothesis that $\beta_1 = \beta_2$, or equivalently, that $\beta_1 - \beta_2 = 0$, has the form

$$t = \frac{b_1 - b_2}{s(b_1 - b_2)}, \qquad (2.21)$$

where $s(b_1 - b_2)$ is the standard deviation of the difference between the two slope coefficients. This test has $n_1 + n_2 - 4$ degrees of freedom. The variance, $s^2(b_1 - b_2)$, of the difference between two slope coefficients is

$$s^2(b_1 - b_2) = \frac{\sum_{i=1}^{n_1}(y_{1i} - \bar{y}_1)^2 + \sum_{i=1}^{n_2}(y_{2i} - \bar{y}_2)^2}{n_1 + n_2 - 4}$$
$$* \left(\frac{1}{\sum_{i=1}^{n_1}(x_{1i} - \bar{x}_1)^2} + \frac{1}{\sum_{i=1}^{n_2}(x_{2i} - \bar{x}_2)^2} \right). \qquad (2.22)$$

When deciding whether or not to reject the null hypothesis that the difference between two regression slopes is zero, one proceeds as follows:

1. Two-sided alternative hypotheses: If $t > t_{\alpha,df}$ then reject H_0.

2. One-sided alternative hypotheses: If $t > t_{\alpha/2,df}$ and $b_1 - b_2$ has the right sign, then reject H_0.

For a data example we use the data in Figure 2.7 again. The thinner regression line for the new sample has a less steep slope than the thicker regression line for the original sample. In the following we answer the question of whether this difference is statistically significant.

The null hypothesis for the statistical test of this question is $\beta_1 = \beta_2$. If this hypothesis prevails, we can conclude that the two samples stem from one population with one regression slope parameter. If the alternative hypothesis prevails, we conclude that the samples stem from different populations with different slope parameters. Table 2.2 contains the data and calculations needed for the comparison of the two samples. (Data and calculations for the original sample are in Table 2.1.) Using these data we calculate the following regression equation for the 1995 sample:

$$\text{Recall}_2 = 97.543 - 6.543 * \text{Trial} + \text{Residual}.$$

We now ask whether the $b_1 = -7.16$ for the original sample is significantly different than the $b_2 = -6.543$ for the present sample. We perform a two-sided test because there is no *a priori* knowledge that guides more specific expectations.

To calculate the variance of the difference between b_1 and b_2 we insert the values from Tables 2.1 and 2.2 into Formula (2.22) and obtain

$$s^2(b_1 - b_2) = \frac{1792.84 + 2056.96}{8 + 8 - 4}\left(\frac{1}{42} + \frac{1}{42}\right) = 15.27.$$

Inserting into (2.21) yields the t statistic

$$t = \frac{-7.16 + 6.54}{\sqrt{15.27}} = -0.158.$$

This test statistic has 8 + 8 - 4 = 12 degrees of freedom. The two-sided tail probability for this t value is $p = 0.877$. This value is greater than $\alpha = 0.05$. Thus, we are in no position to reject the null hypothesis.

2.5.5   More Hypotheses for Slope Coefficients

Sections 2.5.1, 2.5.3, and 2.5.4 presented a selection of tests for the most frequently asked questions concerning slope coefficients in simple regression. As one can imagine, there are many more questions one can ask. We list three sample questions. For details concerning test procedures please consult the cited sources.

First, one can ask whether slope coefficients are constant in the same sample. Consider a repeated measures design where predictors and criteria are repeatedly observed. This design allows researchers to estimate regression parameters separately for each observation point. These estimates can be tested using hierarchical linear models (Bryk & Raudenbush, 1992).

Second, one can ask whether one regression line is consistently located above another, within a given interval of predictor values. There is a t test by Tsutakawa and Hewett (1978) that allows one to answer this question.

Third, one can ask whether the decision to fit a linear regression line is supported by the data. Analysis of variance methods allow one to analyze this question (Neter, Kutner, Nachtsheim, & Wasserman, 1996).

2.5.6 Hypotheses Concerning Intercept Coefficients

The preceding sections introduced readers to methods for testing hypotheses concerning the slope coefficient, β_1. The following three hypotheses were covered in detail: (1) $H_0 : \beta_1 = 0$, (2) $H_0 : \beta_1 = k$, and (3) $H_0 : \beta_1 = \beta_2$. Hypothesis (1) can be shown to be a special case of Hypothesis (2) when we set $k = 0$.

This section presents a t test for testing the hypothesis $H_0 : \beta_0 = k$. As before, we first present a general form for the t test for this hypothesis. This form can be given as

$$t = \frac{b_0 - \beta_0}{s(b_0)}, \tag{2.23}$$

where, as before, b_0 is the estimate of the population parameter, β_0, and $k = \beta_0$. Degrees of freedom are $df = n - 2$. The variance, $s^2(b_0)$, is given by

$$s^2(b_0) = MSE \left(\frac{1}{n} + \frac{\bar{x}^2}{\sum_{i=1}^n (x_i - \bar{x})^2} \right), \tag{2.24}$$

where MSE is the sum of the residuals, divided by $n - 2$, that is, its degrees of freedom. The standard deviation in the denominator of (2.23) is the square root of (2.24).

For a numerical illustration consider again the data in Table 2.2. We test the hypothesis that the estimate, $b_0 = 97.543$, describes a sample that stems from a population with $\beta_0 = 0$. Inserting into (2.24) yields the variance

$$s^2(b_0) = 43.164 \left(\frac{1}{8} + \frac{\left(\frac{36}{8} \right)^2}{42} \right) = 43.164 * 0.607 = 26.207.$$

Inserting into (2.23) yields the t value

$$t = \frac{97.543 - 0}{\sqrt{26.207}} = \frac{97.543}{5.119} = 19.054.$$

The critical two-sided t value for $\alpha = 0.05$ and $df = 6$ is $t = 2.447$. The calculated t is greater than the critical. Thus, we can reject the null hypothesis that $\beta_0 = 0$.

After this example, a word of caution seems in order. The test of whether $\beta_0 = 0$ does not always provide interpretable information. The reason for this caution is that in many data sets researchers do simply not assume that the intercept is zero. As was indicated before, the intercept is equal to the mean of Y only when X was centered before analysis. If this is the case, the above test is equivalent to the t test $H_0 : \mu = 0$.

Again, this test is meaningful only if the observed variable Y can reasonably assumed to have a mean of zero. This applies accordingly when the predictor variable, X, was not centered.

Easier to interpret and often more meaningful is the comparison of the calculated b_0 with some population parameter, β_0. This comparison presupposes that the regression model used to determine both values is the same. For instance, both must be calculated using either centered or noncentered X values. Otherwise, the values are not directly comparable.

2.5.7 Confidence Intervals

The t distribution can be used for both significance testing and estimating confidence intervals. The form of the interval is always

$$\text{lower limit} \leq \beta_j \leq \text{upper limit,}$$

where β_j is the parameter under study. The lower limit of the confidence interval is

$$b_j - s(b_j)\, t_{\alpha/2, n-2},$$

where $t_{\alpha/2, n-2}$ is the usual significance threshold. Accordingly, the upper limit is

$$b_j + s(b_j) t_{\alpha/2, n-2}.$$

For a numerical example consider the slope parameter estimate $b_1 = -6.543$. The 95% confidence interval for this estimate is

$$
\begin{aligned}
-6.543 - 1.014 * 2.447 &\leq \beta_1 \leq -6.543 + 1.014 * 2.447 \\
\Leftrightarrow \qquad -9.024 &\leq \beta_1 \leq -4.062.
\end{aligned}
$$

This confidence interval includes only negative parameter estimates.

This is an illustration of the general rule that if both limits of a confidence interval have the same sign, the parameter is statistically significant.

Chapter 3

MULTIPLE LINEAR REGRESSION

Most typically and frequently, researchers predict outcome variables from more than one predictor variable. For example, researchers may wish to predict performance in school, P, from such variables as intelligence, I, socioeconomic status, SES, gender, G, motivation, M, work habits, W, number of siblings, NS, sibling constellation, SC, pubertal status, PS, and number of hours spent watching TV cartoons, C. Using all these variables simultaneously, one arrives at the following multiple regression equation:

$$\hat{P} = b_0 + b_1 I + b_2 SES + b_3 G + b_4 M + b_5 W + b_6 NS$$
$$+ b_7 SC + b_8 PS + b_9 C. \quad (3.1)$$

This chapter introduces readers to concepts and techniques for multiple regression analysis. Specifically, the following topics are covered: ordinary least squares estimating and testing parameters for multiple regression (Section 3.1) and multiple correlation and determination (Section 3.3).

3.1   Ordinary Least Squares Estimation

The model given in (3.1) is of the form

$$y_i = \beta_0 + \sum_{j=1}^{p} \beta_j x_{ij} + \epsilon_i, \tag{3.2}$$

where j indexes parameters, $p+1$ is the number of parameters to be estimated, p is the number of predictors in the equation, and i indexes cases. The multiple regression model has the same characteristics as the simple regression model. However, the number of predictors in the equation is increased from one to p. Accordingly, the deterministic part of the model must be extended to

$$\beta_0 + \sum_{j=1}^{p} \beta_j x_{ij}.$$

To derive a general OLS solution that allows one to simultaneously estimate all $p + 1$ parameters we use the tools of matrix algebra. In principle the derivation could be achieved without matrix algebra, as was outlined in Section 2.1. We insert the equation for the prediction,

$$\hat{y}_j = b_0 + \sum_{j=1}^{p} b_j x_{ij} + \epsilon_i,$$

into Function (2.7), that is,

$$Q = \sum_{i=1}^{n} (y_i - [b_0 + \sum_{j=1}^{p} b_j x_{ij} + \epsilon_i])^2. \tag{3.3}$$

This is now a function of the $p+1$ b-coefficients. We calculate all partial derivatives of Q with respect to each b-coefficient and set them equal to zero. This results in a system of $p+1$ equations that can be uniquely solved for the $p + 1$ unknown b-coefficients. With three or more b-coefficients, formulas become quickly inconvenient. However, the general strategy remains the same, just the notation is changing. Matrix algebra allows a very compact notation. Therefore, we use matrix notation. For the following derivations we need results of vector differentiation. Appendix C reviews these results.

Before we can apply the rules of vector differentiation we first have to recast Equation (3.3) using matrix algebra. The model given in (3.2) can be written in matrix notation as

$$\mathbf{y} = \mathbf{X}\boldsymbol{\beta} + \boldsymbol{\epsilon},$$

where $\mathbf{y}$ is the vector of observed values, $\mathbf{X}$ is the design matrix, and $\boldsymbol{\epsilon}$ is the residual vector. The design matrix contains the vectors of all predictors in the model. In matrix notation, the OLS criterion to be minimized, Q, is expressed as

$$Q = (\mathbf{y} - \mathbf{Xb})'(\mathbf{y} - \mathbf{Xb}). \tag{3.4}$$

The expressions in the parentheses describe vectors. Transposing a vector and then multiplying it with another vector – in this case with itself – yields the inner vector product. The result of an inner vector product is a scalar, that is, a single number. What we need to find is the vector $\mathbf{b}$ for which this number is the smallest possible.

We now show in detail how one arrives at the well-known least squares solution. The minimization of the sum of squared residuals proceeds in two steps.

1. We determine the first vectorial derivative with respect to $\mathbf{b}$, or, equivalently, we calculate all partial derivatives of Q with respect to the $p + 1$ elements in $\mathbf{b}$.

2. The necessary condition for a minimum of Q at point $\mathbf{b}$ is that the vectorial derivative, that is, all partial derivatives, be zero at that point. The vector $\mathbf{b}$ at which this condition is met can be found by setting this vectorial derivative equal to zero and then solving for $\mathbf{b}$.

We begin with (3.4). Multiplying out yields

$$Q = \mathbf{y}'\mathbf{y} - \mathbf{b}'\mathbf{X}'\mathbf{y} - \mathbf{y}'\mathbf{X}'\mathbf{b} + \mathbf{b}'\mathbf{X}'\mathbf{Xb}. \tag{3.5}$$

To complete the first step of the minimization process we determine the first vectorial derivative of (3.5) using the rules of appendix C. Before we do this, we inspect (3.5). In this expression vector $\mathbf{y}$ and matrix $\mathbf{X}$ are known. Only the parameter vector, $\mathbf{b}$, is unknown. Therefore, we need to determine the derivative in regard to $\mathbf{b}$.

Because the vectorial derivative of a sum is equal to the sum of the vectorial derivatives, we apply the rules from the Appendix separately to the four terms of (3.5), that is,

$$\frac{\partial}{\partial \mathbf{b}}Q = \frac{\partial}{\partial \mathbf{b}}\mathbf{y}'\mathbf{y} - \frac{\partial}{\partial \mathbf{b}}\mathbf{b}'\mathbf{X}'\mathbf{y} - \frac{\partial}{\partial \mathbf{b}}\mathbf{y}'\mathbf{X}'\mathbf{b} + \frac{\partial}{\partial \mathbf{b}}\mathbf{b}'\mathbf{X}'\mathbf{X}\mathbf{b}. \tag{3.6}$$

For the first term in (3.6) we apply Rule 3, (Appendix C) for the second and third terms we apply Rule 2 and Rule 1, respectively, and for the fourth term we use Rule 5. We obtain

$$\frac{\partial}{\partial \mathbf{b}}Q = 0 - \mathbf{X}'\mathbf{y} - \mathbf{X}'\mathbf{y} + [\mathbf{X}'\mathbf{X} + (\mathbf{X}'\mathbf{X})']\mathbf{b}. \tag{3.7}$$

Because $(\mathbf{X}'\mathbf{X})' = \mathbf{X}'\mathbf{X}$ we can simplify (3.7) and obtain

$$\frac{\partial}{\partial \mathbf{b}}Q = -2\mathbf{X}'\mathbf{y} + 2\mathbf{X}'\mathbf{X}\mathbf{b}. \tag{3.8}$$

However, (3.8) does not provide us yet with the estimate for the parameter vector $\mathbf{b}$ that yields the smallest possible sum of the squared residuals. To obtain this vector, we need to perform the second step of the minimization process. This step involves setting (3.8) equal to zero and transforming the result so that $\mathbf{b}$ appears on one side of the equation and all the other terms on the other side. Setting (3.8) equal to zero yields

$$-2\mathbf{X}'\mathbf{y} + 2\mathbf{X}'\mathbf{X}\mathbf{b} = 0. \tag{3.9}$$

Dividing all terms of (3.9) by 2 and moving the first term to the right-hand side of the equation yields what is known as the *normal equations*

$$\mathbf{X}'\mathbf{X}\mathbf{b} = \mathbf{X}'\mathbf{y}. \tag{3.10}$$

To move the term $\mathbf{X}'\mathbf{X}$ to the right-hand side of the equation, we premultiply both sides of the equation with $(\mathbf{X}'\mathbf{X})^{-1}$, that is, with the inverse of $\mathbf{X'X}$. This yields

$$(\mathbf{X'X})^{-1}(\mathbf{X'X})\mathbf{b} = (\mathbf{X'X})^{-1}\mathbf{X'y},$$

and because $(\mathbf{X'X})^{-1}(\mathbf{X'X}) = \mathbf{I}$, we obtain

$$\mathbf{b} = (\mathbf{X'X})^{-1}\mathbf{X'y}. \tag{3.11}$$

Equation (3.11) is the ordinary least squares solution for parameter vector $\mathbf{b}$. Recall from Appendix A on matrix algebra that not all matrices have an inverse. That the inverse of $\mathbf{X'X}$ exists follows from the assumption that $\mathbf{X}$ is of full column rank made in Section 2.5. Vector $\mathbf{b}$ contains all $p+1$ parameter estimates for the multiple regression problem, that is, the intercept and the coefficients for the variables.

To illustrate the meaning of the coefficients in multiple regression, consider the case where $p+1$ independent variables, X_i, along with their coefficients are used to explain the criterion, Y. The first of these coefficients, that is, b_0, is, as in simple regression, the intercept. This is the value that Y assumes when all independent variables are zero. As was explained in the chapter on simple regression, this value equals the mean of Y, $\bar{y}$, if all variables X_j are centered (for $j = 1, ..., p$).

The remaining p coefficients constitute what is termed a regression surface or a response surface. This term is appropriate when there are two X variables, because two variables indeed span a surface or plane. When there are three or more X variables, the term "regression hyperplane" may be more suitable. To illustrate, consider the following multiple regression equation, where we assume the regression coefficients to be known:

$$\hat{Y} = 14 + 0.83X_1 - 0.34X_2.$$

When $X_1 = X_2 = 0$, the criterion is $\hat{Y} = 14$. The independent variables, X_1 and X_2, span the regression surface given in Figure 3.1. The grid in Figure 3.1 depicts the regression surface for $-1.0 \leq X_1, X_2 \leq +1.0$. The regression coefficients can be interpreted as follows:

- Keeping X_2 constant, each one-unit increase in X_1 results in a 0.83-unit increase in Y; this change is expressed in Figure 3.1 by the grid

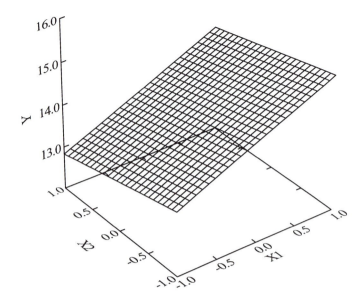

Figure 3.1: Regression hyperplane $\hat{y} = 14 + 0.83x_1 - 0.34x_2$.

lines that run parallel to the X_1 axis.

- Keeping X_1 constant, each one-unit increase in X_2 results in a 0.34-unit decrease in Y; this change is expressed in Figure 3.1 by the grid lines that run parallel to the X_2 axis.

In more general terms, parameters in multiple regression indicate the amount of change in Y that results from a given variable when all other variables are kept constant. Therefore, coefficients in multiple regression are often called *partial regression coefficients*. The change is expressed in units of Y.

When a multiple regression equation involves more than two predictors, the regression hyperplane cannot be depicted as in Figure 3.1. Consider an equation with four predictors. The hyperplane spanned by four predictors has four dimensions. The entire data space under study has five dimensions and cannot be graphed.

One important characteristic of the regression surface depicted in Figure 3.1 is that it is completely linear. Linear regression lines and surfaces are valid if there is no interaction between predictors. In this case, regression predictors are said to have *additive effects*. If, however, there exist interactions, the regression surface is no longer linear. It may be curved or evince leaps. As a matter of course, this can also be the case if the relation between predictor and criterion is basically nonlinear.

To test whether the parameters estimated in multiple regression are significantly different than zero, researchers typically employ either the t test or the F test. The F test will be explained later in this chapter. The t test can be specified in the following two steps:

1. Estimation of the covariance matrix of the **b** vector.

2. Division of b_k by the kth diagonal element of the estimated covariance matrix. These two steps result in a t test that has the well-known form

$$t = \frac{b_k}{s(b_k)},\tag{3.12}$$

where b_k is the kth estimate of the coefficient of the multiple regression model, and $s(b_k)$ is the estimate for the standard deviation of this coefficient.

To derive the covariance matrix of the **b** vector we use the following result:

$$\mathbf{C}(\mathbf{Ay}) = \mathbf{AC}(\mathbf{y})\mathbf{A}'.$$

Recall from Section 2.2.2 that $\mathbf{C}(\mathbf{y}) = \sigma^2\mathbf{I}$, which expresses the assumptions that the observations are independent and have constant variance.

The covariance matrix of the **b** vector, $\mathbf{C}(\mathbf{b})$, can then be calculated as follows:

$$
\begin{aligned}
\mathbf{C}(\mathbf{b}) &= \mathbf{C}[(\mathbf{X'X})^{-1}\mathbf{X'y}] \\
&= (\mathbf{X'X})^{-1}\mathbf{X'C}(\mathbf{y})\mathbf{X}(\mathbf{X'X})^{-1} \\
&= (\mathbf{X'X})^{-1}\mathbf{X'}\sigma^2\mathbf{IX}(\mathbf{X'X})^{-1} \\
&= \sigma^2(\mathbf{X'X})^{-1}\mathbf{X'X}(\mathbf{X'X})^{-1} \\
&= \sigma^2(\mathbf{X'X})^{-1}.
\end{aligned}
$$

These equations give the *true* covariance matrix. When analyzing real life data, however, we rarely know the population variance, σ^2. Thus, we substitute the mean squared error, *MSE*, for σ^2, and obtain

$$\hat{\mathbf{C}}(\mathbf{b}) = \text{MSE} \; (\mathbf{X}'\mathbf{X})^{-1}, \qquad (3.13)$$

where the hat indicates that this is an estimate of the true covariance matrix. The mean squared error,

$$\text{MSE} = \frac{1}{n-p} \sum_{i=1}^{n} e_i,$$

is an unbiased estimate of the error variance σ^2. Taking the square roots of the diagonal elements of the covariance matrix given in (3.13) yields the estimates for the standard deviations used in the denominator in (3.12). These are the values that typically appear in regression analysis output tables as *standard error* of b_k.

3.2   Data Example

The following data example uses a classical data set, Fisher's iris data. The set contains information on $n = 150$ iris flowers. The following five variables were observed: Species (three categories), Sepal Length, Sepal Width, Petal Length, and Petal Width. To illustrate multiple regression with continuous variables, we predict Petal Length from Petal Width and Sepal Width. We use all cases that is, we do not estimate parameters by Species.[1] The distribution of cases appears in Figure 3.2, coded by Species.

We estimate parameters for the following regression equation:

Petal Length $= b_0 + b_1 *$ Petal Width $+ b_2 *$ Sepal Width $+$ Residual.

---

[1] It should be noted that disregarding Species as a possibly powerful predictor could, when analyzing real data, result in *omission bias*. This is the bias caused by the absence of powerful predictors in a regression model. In the present context, however, we have to omit predictor Species, because multicategorical predictors will be covered later, in Chapter 4.

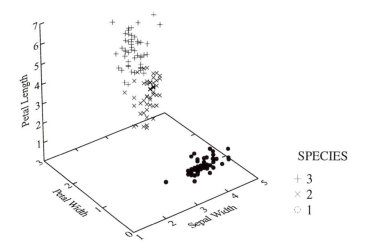

Figure 3.2: Three species of iris flowers.

OLS multiple regression yields the following parameter estimates

Petal Length $= 2.26 + 2.16 *$ Petal Width

$$- 0.35 * \text{Sepal Width} + \text{Residual}.$$

Figure 3.3 depicts the regression plane created by this function.

To determine whether the regression coefficients differ from zero we use the t test as explained above. Having already obtained the parameter estimates as given in the last equation, all we need to be able perform the t test is the estimated standard error of each estimated regression coefficient, that is, the square roots of the diagonal elements of $\text{MSE}(\mathbf{X'X})^{-1}$. The design matrix $\mathbf{X}$ has 150 rows and 3 columns; therefore $\mathbf{X'X}$ is a square matrix with 3 rows and columns. In our example the inverse of $\mathbf{X'X}$ yields

$$(\mathbf{X'X})^{-1} = \begin{pmatrix} 0.470 & -0.042 & -0.135 \\ -0.042 & 0.013 & 0.009 \\ -0.135 & 0.009 & 0.041 \end{pmatrix}.$$

Note that this matrix is symmetric. For these data, $MSE = 0.209$.

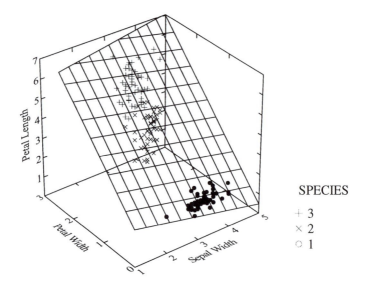

Figure 3.3: Regression surface for prediction of iris petal length.

Now, multiplying $(\mathbf{X'X})^{-1}$ by MSE yields the estimated covariance matrix of the estimated parameter vector.

$$\hat{\mathbf{C}}(\mathbf{b}) = 0.209\ (\mathbf{X'X})^{-1} = \begin{pmatrix} 0.0983 & -0.0088 & -0.0282 \\ -0.0088 & 0.0028 & 0.0018 \\ -0.0282 & 0.0018 & 0.0085 \end{pmatrix}.$$

The diagonal elements of this matrix are the estimated variances of the estimated regression coefficients. The off-diagonal elements contain the covariances. The first diagonal element is the estimated variance of the intercept estimate, the second the estimated variance of the Petal Width coefficient, and the third the estimated variance of the Sepal Width coefficient. Because we do not need the variance estimates, but rather the corresponding standard errors, we take the square roots of the three diagonal elements. The coefficients with standard errors in parentheses are

Intercept	2.26	(0.313)
Petal Width	2.16	(0.053)
Sepal Width	-0.35	(0.092).

We finally perform the t test, by simply dividing the estimate of the regression coefficient by its estimated standard error. Since the intercept parameter is not of interest, we do not consider testing it.

Both slope parameters are statistically significantly greater than zero. More specifically, we obtain $t_{147} = 40.804$ and $t_{147} = -3.843$ for b_1 and b_2, respectively. The two-tailed probabilities for both parameters are $p < 0.01$. Both t tests have 147 degrees of freedom. For these t tests the degrees of freedom are given by the formula $df = n - p - 1$, where n is the number of observations and p is the number of predictor variables. (For significance testing in multiple regression using the F test see Section 3.4.) Computer programs for regression analysis usually report in their output of regression analyses the estimated regression coefficients, the corresponding standard errors, the t values, and the p values. Often, the whole covariance matrix of the estimated regression coefficients is not needed. Therefore, it is typically not printed by default, but can often be requested as an output option.

Many researchers ask whether, beyond statistical significance, a good portion of variance of the criterion can be accounted for by the predictors. This portion of variance can be measured via the square of the multiple correlation coefficient. The next chapter covers the multiple R.

3.3 Multiple Correlation and Determination

To explain the concept of multiple correlation we use centered variables. Centering is a linear transformation that involves subtracting the arithmetic mean from each value. The basic equation that underlies the concept of multiple correlation and determination is that of decomposition of variability:

$$\text{Total Variability} = \text{Residual Variability} + \text{Explained Variability.} \quad (3.14)$$

The total variability, also termed variability of the criterion, can be expressed as follows:

$$\text{Total Variability} = \mathbf{y}'\mathbf{y}. \tag{3.15}$$

For the sake of simplicity, and because we do not need this detail for the following explanation, we express the residual variability as the Sum of Squared Residuals, *SSR*,

$$\text{Residual Variability} = \text{SSR}. \tag{3.16}$$

In a fashion analogous to the total variability we express the explained variability as

$$\text{Explained Variability} = \hat{\mathbf{y}}'\hat{\mathbf{y}}. \tag{3.17}$$

Using (3.15), (3.16), and (3.17) we can reexpress the decomposition Formula (3.14) as

$$\mathbf{y}'\mathbf{y} = \text{SSR} + \hat{\mathbf{y}}'\hat{\mathbf{y}}. \tag{3.18}$$

Dividing (3.18) by the criterion variability expresses variability components as proportions of unity:

$$\frac{\mathbf{y}'\mathbf{y}}{\mathbf{y}'\mathbf{y}} = \frac{\text{SSR}}{\mathbf{y}'\mathbf{y}} + \frac{\hat{\mathbf{y}}'\hat{\mathbf{y}}}{\mathbf{y}'\mathbf{y}}$$

or

$$1 = \frac{\text{SSR}}{\mathbf{y}'\mathbf{y}} + \frac{\hat{\mathbf{y}}'\hat{\mathbf{y}}}{\mathbf{y}'\mathbf{y}}. \tag{3.19}$$

Equation (3.19) can be read as follows: The criterion variability, expressed as 1, is composed of one portion that remains unexplained, the residual variability, and one portion that is explained. The latter portion appears in the second term on the right-hand side of (3.19). This term is the ratio of explained variability to criterion variability. It is termed the coefficient of multiple determination, R^2.

In other words,

$$R^2 = \frac{\hat{\mathbf{y}}'\hat{\mathbf{y}}}{\mathbf{y}'\mathbf{y}},$$

and

$$R = \sqrt{\frac{\hat{\mathbf{y}}'\hat{\mathbf{y}}}{\mathbf{y}'\mathbf{y}}}.$$

Because of

$$\hat{\mathbf{y}}'\hat{\mathbf{y}} \leq \mathbf{y}'\mathbf{y},$$

we have always $0 \leq R^2 \leq 1$ and, accordingly, $0 \leq R \leq 1$.

In general, the coefficient of multiple determination is a measure of the strength of the association between one criterion and multiple predictors.

In the iris flowers data example of the last section we predicted Petal Length from Petal Width and Sepal Width. This predictor set allowed us to explain $R^2 = 0.933$, that is, 93.3% of the variance of Petal Length. The multiple correlation was $R = 0.966$.

3.3.1 Expectancy of R^2 and Significance Testing

There exist statistical significance tests that allow one to test the null hypothesis

$$H_0 : \rho^2 = 0, \tag{3.20}$$

where ρ is the multiple correlation in the population. The F statistic for this null hypothesis is

$$F = \frac{R^2(n - p - 1)}{(1 - R^2)p}, \tag{3.21}$$

where p denotes the number of predictors and n is the sample size; p and $n - p - 1$ are the degrees of freedom for the F statistic.

Taking up the iris flower example from Section 3.2, we insert the estimate for the coefficient of multiple determination into (3.21) and obtain

$$F = \frac{0.933(150 - 2 - 1)}{(1 - 0.933)2} = \frac{137.15}{0.067} = 1023.52,$$

a value that is statistically significant.

While appropriate in many instances, the null hypothesis given in (3.20) may not always be reasonable to ask (Huberty, 1994). The reason for this caution is that the expectancy of R^2, $E(R^2)$, that is, the long-run mean of R^2, is greater than zero if $\rho = 0$ (Morrison, 1990). Specifically,

$$E(R^2) = \frac{p}{n - 1}, \quad \text{if } \rho = 0.$$

For example, in a sample of $n = 26$ with $p = 6$ parameters to estimate we have $E(R^2) = 5/25 = 0.20$. This means that, across many independent samples, one can expect to obtain, by chance, $R^2 = 0.20$. Thus, it can occur that a R^2 is statistically greater than zero but is hardly different or even smaller than what one can expect it to be from chance.

Therefore, it has been proposed to replace the null hypothesis $H_0 : \rho^2 = 0$ by

$$H_0 : \rho^2 = \rho_0^2,$$

where $\rho_0^2 = E(R^2)$. This null hypothesis can be tested using the F statistic

$$F = \frac{R^2(n - p - 1)^2}{(1 - R^2)p(2n - p - 2)}, \tag{3.22}$$

with numerator degrees of freedom

$$df_1 = \frac{4p(n - 1)}{3n + p - 3} \tag{3.23}$$

and denominator degrees of freedom $df_2 = n - p - 1$ (Darlington, 1990; Huberty, 1994).

Consider the data example with $n = 26$ and $p = 6$. Suppose we have an estimated $R^2 = 0.50$. By inserting into (3.22) we obtain

$$F = \frac{0.5(26 - 6 - 1)^2}{(1 - 0.5)(6 - 1)(2 * 26 - 6 - 2)} = \frac{180.5}{110} = 1.64,$$

with numerator degrees of freedom

$$df_1 = \frac{4 * 6(26 - 1)}{3 * 26 + 5 - 3} = \frac{600}{81} = 7.41.$$

Denominator degrees of freedom are $df_2 = 26 - 6 - 1 = 19$. With these degrees of freedom the F value has a tail probability of $p = 0.18$. Thus, we have no reason to reject the null hypothesis that $\rho^2 = 0.20$.

Now, in contrast to this decision, we calculate $F = 3.17$ from inserting into (3.21). With $df_1 = 6$ and $df_2 = 19$, this value has a tail probability of $p = 0.0251$. Thus, we would reject the null hypothesis that $\rho^2 = 0$.

When interpreting results, we now have a complicated situation: while ρ^2 is greater than zero, it is not greater than what one would expect from chance. In other words, ρ^2 is greater than some value that is smaller than what one would expect from chance. Since this smaller value may be arbitrary, it is recommended using the test statistic given in (3.22). It allows one to test whether ρ^2 is greater than one would expect in the long run for a sample of size n and for p predictors.

In contrast (see Huberty, 1994, p. 353), the null hypothesis $H_0 : R^2 = 0$ is equivalent to the test that *all* correlations between the predictors and the criterion are *simultaneously* equal to 0. Thus, the F test for this hypothesis can be applied in this sense.

The paper by Huberty (1994) has stimulated a renewed discussion of methods for testing hypotheses concerning R^2. Snijders (1996) discusses an estimate of ρ^2 that goes back to Olkin and Pratt (1958). This estimate is very hard to calculate. A second-order approximation of this estimator is

$$R^2_{OP2} = 1 - \frac{n - 3}{n - p - 1}(1 - R^2)$$
$$* \left\{ 1 + \frac{1}{n - p - 1}(1 - R^2) + \frac{8}{(n - p - 1)(n - p - 3)}(1 - R^2) \right\}.$$

As long as $n < 50$, this estimator has considerably less bias than the ordinary R^2.

Yung (1996) notes that Huberty's (1994) numerator degrees of freedom formula [Formula (3.23)] is incorrect. The correct formula is

$$\text{df} = \frac{[(n-1)a + p]^2}{(n-1)(a+2)a + p} ,$$

where $a = \rho^2/(1 - \rho^2)$.

In addition Yung (1996) notes that the test proposed by Huberty is inconsistent, and cites Stuart and Ord (1991) with a more general formula for the expected value of R^2, given $\rho^2 = k$. The formula is

$$E(R^2|\rho^2 = k) = k + \frac{p}{n-1}(1-k) - \frac{2(n-p-1)}{n^2-1}k(1-k) + \frac{a}{n^2}.$$

The last term in this formula vanishes when $\rho^2 = 0$ or when $\rho^2 = 1$. Most important for practical purposes is that Yung's results suggest that a test of the null hypothesis that the observed R^2 is statistically not different than some chance value can proceed just as the above classical significance test. Only the significance level α must be adjusted to accommodate the probability of the chance score. If the classical test is performed, for instance, in a sample of $n = 30$ with $p = 5$ predictors, and the population $\rho^2 = 0$ and the specified $\alpha^* = 0.1$, then the appropriate significance level is $\alpha = 0.04$. Yung (1996, p.296) provides figures for $\rho^2 = 0, 0.15, 0.30$, and 0.90 that cover between 1 and 8 predictors, and give the appropriate significance thresholds for the sample sizes $n = 10, 11, 14, 17, 20, 30, 50$, and 100. More elaborate tables need to be developed, and the test needs to be incorporated into statistical software packages.

3.4   Significance Testing

Section 2.5.1 presented an F test suitable for simple regression. This chapter presents a more general version of this test. The test requires the same conditions to be met as listed in Section 2.5.

The F test to be introduced here involves comparing two nested models, the *constrained model* and the *unconstrained model*. The unconstrained model involves all predictors of interest. Consider a researcher that is interested in predicting performance in school from six variables. The unconstrained model would then involve all six variables.

To measure whether the contribution made by a particular variable is statistically significant, the researcher eliminates this variable. The resulting model is the constrained model. If this elimination leads to statistically significant reduction of explained variance, the contribution of this variable is statistically significant. This applies accordingly when more than one variable is eliminated. This way one can assess the contributions not only of single variables but also of entire variable groups.

Consider the following example. As is well known, intelligence tests are made of several subtests, each one yielding a single test score. Using different subtests researchers try to separate, say, verbal intelligence from abstract and more formal mental abilities. Thus, it is often of interest to assess the contribution made by a group of intelligence variables to predicting, for instance, performance in school.

To assess the contribution of the constrained model relative to the unconstrained model, we compare the R^2 values of these two models. Let R_u^2 denote the portion of variance accounted for by the unconstrained model, and R_c^2 the portion of variance accounted for by the constrained model. Then, the F statistic for comparing the constrained with the unconstrained models is

$$F = \frac{(R_u^2 - R_c^2)/(p_u - p_c)}{(1 - R_u^2)/(n - p_u - 2)}, \quad \text{for} \quad R_u^2 < 1. \tag{3.24}$$

The upper numerator of (3.24) subtracts the portions of variance accounted for by the unconstrained and the constrained models from each other. What remains is the portion of variance accounted for by the eliminated variable(s). The upper denominator contains the number of parameters estimated by the eliminated variable(s). In the lower numerator we find the portion of variance that remains unaccounted for by the unconstrained model, weighted by the degrees of freedom for this model.

The upper denominator in (3.24) contains the numerator degrees of freedom. The bottom denominator contains the denominator degrees of freedom.

The null hypothesis tested with (3.24) is

$$H_0 : \beta = 0$$

or, in words, the weight of the eliminated variables is zero in the population under study. The alternative hypothesis is

$$H_0 : \beta \neq 0$$

or, in words, the weight of the eliminated variables is unequal to zero.

Note that this test is applicable in situations where β is a vector. This includes the case that β has length one, or equivalently, that β is a real number. For the latter case, that is, when there is only one regression coefficient to be tested, we already have described the t test that can be used. It will be illustrated in the exercises that the t test and the F test just described are equivalent when β is one real number. This equivalence between the two tests was already noted in Section 2.5.2 where significance tests for the simple linear regression were illustrated. The F test given above can be considered a generalization of the t test.

In the following data example we attempt to predict Dimensionality of Cognitive Complexity, *CC1*, from the three predictors Depth of Cognitive Complexity, *CC2*, Overlap of Categories, *OVC*, and Educational Level, *EDUC*. Dimensionality measures the number of categories an individual uses to structure his or her mental world. Depth of Cognitive Complexity measures the number of concepts used to define a category of Cognitive Complexity. Overlap of Categories measures the average number of concepts two categories share in common. Educational Level is measured by terminal degree of formal school training, a proxy for number of years of formal training. A sample of $n = 327$ individuals from the adult age groups of 20 – 30, 40 – 50, and 60 and older provided information on Educational Level and completed the cognitive complexity task.

Using these data we illustrate two types of applications for the F test given in (3.24). The first is to determine whether the β for a given single predictor is statistically significantly different than zero. The second application concerns groups of predictors. We test whether two predictors as a group make a statistically significant contribution.

The first step of data analysis requires estimation of the unconstrained model. We insert all three predictors in the multiple regression equation. The following parameter estimates result:

$$CC1 = 22.39 - 0.21 * CC2 - 22.40 * OVC + 0.39 * EDUC + \text{Residual}.$$

All predictors made statistically significant contributions (one-tailed tests), and the multiple $R^2 = 0.710$ suggests that we explain a sizeable portion of the criterion variance. Readers are invited to test whether this R^2 is greater than what one would expect from chance.

To illustrate the use of the F test when determining whether a single predictor makes a statistically significant contribution, we recalculate the equation omitting *EDUC*. We obtain the following parameter estimates:

$$CC1 = 31.40 - 0.21 * CC2 - 22.83 * OVC + \text{Residual}.$$

Both predictors make statistically significant contributions, and we calculate a multiple $R^2 = 0.706$.

To test whether the drop from $R^2 = 0.710$ to $R^2 = 0.706$ is statistically significant we insert in (3.24),

$$F = \frac{\frac{0.7097 - 0.7064}{4 - 3}}{\frac{1 - 0.7097}{327 - 4 - 2}} = 3.6490.$$

Degrees of freedom are $df_1 = 1$ and $df_2 = 321$. The tail probability for this F value is $p = 0.0570$. The t value for predictor *EDUC* in the unconstrained model was $t = 1.908$ ($p = 0.0573$; two-sided), and $t^2 = 1.908^2 = 3.6405 = F$. The small difference between the two F values is due to rounding. Thus, we can conclude that predictor *EDUC* makes a statistically significant contribution to predicting *CC1*. However, the value added by *EDUC* is no more than 0.4 % of the criterion variance.

One more word concerning the equivalence of the t test and the F test. The t distribution, just like the normal, is bell-shaped and symmetrical. Thus, the t test can, by only using one half of the distribution, be performed as a one-sided test. In contrast, the F distribution is asymmetric. It may be the only sampling distribution that never approximates the normal, even when sample sizes are gigantic. Thus, there is no one-sided version of the F test.

Information concerning the contribution of single predictors is part of standard computer printouts.[2] In addition, one also finds the F test that contrasts the unconstrained model with the model that only involves the intercept. However, one rarely finds the option to assess the contribution made by groups of predictors. This option is illustrated in the following example. In the example we ask whether the two cognitive complexity variables, Depth and Overlap, as a group make a statistically significant contribution to explaining Breadth. This is equivalent to asking whether these two predictors *combined* make a significant distribution. It is possible, that groups of predictors, none of which individually makes a significant contribution, do contribute significantly. To answer this question we first calculate the parameters for the simple regression that includes only Education as predictor. The resulting equation is

$$CC1 = 1.963 + 1.937 * EDUC + \text{Residual}.$$

The β for predictor Education is statistically significant ($t = 5.686; p < 0.01$), and $R^2 = 0.090$. Inserting into (3.24) yields

$$F = \frac{\frac{0.710 - 0.090}{4 - 2}}{\frac{1 - 0.710}{327 - 4 - 2}} = \frac{0.31}{0.0009} = 343.15.$$

Degrees of freedom are $df_1 = 2$ and $df_2 = 321$. The tail probability for this F value is $p < 0.01$. Thus, we can conclude that the two variables of cognitive complexity, *CC2* and *OVC*, as a group make a statistically significant contribution to predicting *CC1*.

As an aside it should be noted that the F test that tests whether the β for a group of variables is unequal to zero can be applied in analysis of variance to test the so-called main effects. Each main effect involves one (for two-level factors) or more (for multiple-level factors) coding vectors in the design matrix, $\mathbf{X}$. Testing all coefficients together that describe one factor yields a test for the main effect. The same applies when testing together all vectors that describe an interaction. Testing single vectors is equivalent to testing single contrasts.

---

[2]What we discuss here is standard output in SYSTAT's MGLH module and in the Type III Sum of Squares in the SAS-GLM module. Variable groups can be specified in SYSTAT.

Chapter 4

CATEGORICAL PREDICTORS

We noted in Section 2.2.1 that the linear regression model does not place constraints on the predictor, X, that would require X to be distributed according to some sampling distribution. In addition, we noted that X can be categorical. The only constraint placed on X was that it be a constant, that is, a measure without error.

In the present chapter we discuss how to perform simple regression analysis using categorical predictors. Specifically, we discuss regression analysis with the predictor measured at the nominal level. Example of such predictors include religious denominations, car brands, and personality types. An example of a two-category predictor is gender of respondents. This chapter focuses on two-category predictors. Section 4.2 covers multiple-category predictors. In each case, the scale that underlies X does not possess numerical properties that would allow one to perform operations beyond stating whether two objects are equal. Therefore, the step on X that determines the regression slope is not quantitatively defined. Not even the order of categories of X is defined. The order of categories can be changed arbitrarily without changing the information carried by a categorical variable.

However, there are means for using categorical, nominal-level predictors in regression analysis. One cannot use categorical predictors in the

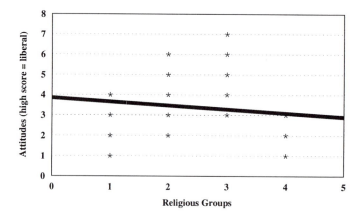

Figure 4.1: Perceived attitudes toward issues in education in four religious groups.

usual way, that is, inserting it in the regression equation "as is." Rather, one must decompose categorical predictors. This decomposition involves creating dummy variables or effect coding variables that contrast categories or groups of categories.

Consider the following example. A researcher asks whether adolescents view faith as predicting liberality of attitudes toward key issues of education. Adolescents from the following four groups responded to a questionnaire: Roman Catholic, Protestant, Agnostics, and Muslims. The Catholics were assigned a 1, the Protestants were assigned a 2, the Agnostics were assigned a 3, and the Muslims were assigned a 4. These numbers must be interpreted at the nominal scale level, that is, they only serve to distinguish between these four groups. There is no ranking or interpretation of intervals. Any other assignment of numbers would have served the same purpose, as long as the numbers are different from each other. Assigning symbols or characters is an alternative to assigning numbers.

The response scale, Y, ranged from 0 to 7, with 0 indicating the least liberal attitudes and 7 indicating the most liberal attitudes. Figure 4.1 displays the responses of the $N = 40$ adolescents (overlapping values are indicated as one value).

The figure suggests that Agnostics are viewed as displaying the most

Table 4.1: *Means and Standard Errors of Responses to Attitude Questionnaire*

Group of Respondents	Descriptive Statistics		
	Mean	Standard Dev.	N
Catholics (1)	2.7	0.95	10
Protestants (2)	4.1	1.20	10
Agnostics (3)	4.9	2.10	10
Muslims (4)	1.8	0.79	10

liberal attitudes. Muslims are viewed as displaying the least liberal attitudes. Protestants and Catholics are in between. Table 4.1 presents means and standard deviations for the four groups of respondents.

Figure 4.1 also displays a regression line. *This line is incorrect!* It was estimated under the assumption that the numerical values assigned to the four respondent groups operate at the interval level. However, the numerical values simply serve as names of the respondent groups. Thus, they operate at the nominal level and are arbitrary. Readers are invited to assign the "1" to the Muslims, to increase all other group indicators by 1, and to recalculate the regression line. The result, a regression line with a positive slope, seemingly contradicts the result presented in Figure 4.1. This contradiction, however, is irrelevant, because *both solutions are false.*[1]

4.1 Dummy and Effect Coding

There are several equivalent ways to create dummy variables and effect coding variables for analysis of variance and regression analysis. In the present section we introduce two of these methods.

A dummy variable (also termed indicator variable or binary variable) can only assume the two values 0 and 1. Dummy variables can be used to discriminate between the categories of a predictor. Consider the following

---

[1]The calculated regression equation for the line presented in Figure 4.1 is: Attitude $= 0.38 - 0.19 *$ Group + Residual, and $R^2 = 0.017$; the calculated regression line for the recoded group values is: Attitude $= 0.25 + 0.35 *$ Group + Residual, and $R^2 = 0.059$.

example. A researcher is interested in predicting durability of cars from car type. Two types of cars are considered, pickup trucks and sedans. The researcher sets up a dummy variable as follows:

$$X_1 = \begin{cases} 1 & \text{if car is a pickup} \\ 0 & \text{if car is a sedan} \end{cases} \tag{4.1}$$

and

$$X_2 = \begin{cases} 1 & \text{if car is a sedan} \\ 0 & \text{if car is a pickup.} \end{cases} \tag{4.2}$$

The constant in the design matrix, $\mathbf{X}$, is often also considered a dummy variable (although it only assumes the value 1).

While intuitively plausible, the approach demonstrated using Formulas (4.1) and (4.2) leads to a problem. The problem is that the design matrix, $\mathbf{X}$, will have linearly dependent columns and, therefore, not have full column rank. Recall that in Section 2.2.1 we assumed that the design matrix was of full column rank, and that this can always be achieved by reparameterization. Consider the following example. The researcher investigating durability of cars analyzes data from the first five cars: three pickups and two sedans. The design matrix, $\mathbf{X}$, for these five cars, appears below:

$$\mathbf{X} = \begin{pmatrix} 1 & 1 & 0 \\ 1 & 1 & 0 \\ 1 & 1 & 0 \\ 1 & 0 & 1 \\ 1 & 0 & 1 \end{pmatrix}. \tag{4.3}$$

In this matrix, the first column is the sum of the last two columns. Thus, $\mathbf{X}$ is not of full column rank. As a result, the product $\mathbf{X'X}$ is singular also, and there are no unique estimates of regression coefficients.

The problem is solved by *reparameterization*. The term reparameterization indicates that the meaning of the parameters depends on how the rank deficiency of $\mathbf{X}$ is overcome. There are several ways to achieve this. The most obvious, simple, and frequently used option involves dropping one of the vectors of the indicator variable. In the car-type example, dropping the second vector results in the following design matrix:

$$\mathbf{X} = \begin{pmatrix} 1 & 1 \\ 1 & 1 \\ 1 & 1 \\ 1 & 0 \\ 1 & 0 \end{pmatrix}.$$

In contrast to (4.3), this matrix is easily analyzed using standard simple regression because now the two columns of $\mathbf{X}$ are linearly independent and, therefore, $\mathbf{X}$ is of full column rank. In general, for a variable with c categories, one sets up no more than $c - 1$ independent dummy coding vectors. The following data example analyzes data that describe the length of usage of three pickup trucks and two sedans, measured in years. The regression equation for these data can be presented in the following form:

$$\begin{pmatrix} 6 \\ 10 \\ 8 \\ 9 \\ 10 \end{pmatrix} = \begin{pmatrix} 1 & 1 \\ 1 & 1 \\ 1 & 1 \\ 1 & 0 \\ 1 & 0 \end{pmatrix} \begin{pmatrix} b_0 \\ b_1 \end{pmatrix} + \begin{pmatrix} e_1 \\ e_2 \\ e_3 \\ e_4 \\ e_5 \end{pmatrix}.$$

Estimating the parameters for this equation yields

$$\text{Years} = 9.5 - 1.5 * \text{Type of Car} + \text{Residual}.$$

The coefficient of determination is $R^2 = 0.24$ and the t test for the slope coefficient is not significant ($t = -0.0976; p = 0.401$). That the slope coefficient is not statistically different from zero is most probably due to the small sample size and the resulting lack of power.

The interpretation of the estimated slope coefficient is that for pickup trucks length of usage is about one and a half years shorter than that for sedans. To be more specific, we estimate length of usage in years for pickup trucks to be $9.5 - 1.5 * 1 = 8$ years, while for sedans we calculate estimated length of usage to be $9.5 - 1.5 * 0 = 9.5$ years. The estimated difference is therefore 1.5 years. The intercept parameter estimates the mean value for sedans, and the slope coefficient estimates the difference of mean length of usage between sedans and pickup trucks.

As an alternative to dummy coding, effect coding is often used. Effect coding makes it easy to set up comparisons. Comparisons always involve two groups of cases (subjects, data carriers, respondents, etc.). As this section focuses on two-category predictors, effect coding is established quite easily. Vectors are set up according to the following rule: Members of the first group are assigned a 1, and members of the second group are assigned a -1. It is of no importance which of the groups is assigned the 1 and which is assigned the -1. Reversing assignment results in a change in the sign of the regression coefficient. Results of significance testing remain the same.

Using effect coding, the regression equation is set up as follows:

$$
\begin{pmatrix} 6 \\ 10 \\ 8 \\ 9 \\ 10 \end{pmatrix} = \begin{pmatrix} 1 & 1 \\ 1 & 1 \\ 1 & 1 \\ 1 & -1 \\ 1 & -1 \end{pmatrix} \begin{pmatrix} b_0 \\ b_1 \end{pmatrix} + \begin{pmatrix} e_1 \\ e_2 \\ e_3 \\ e_4 \\ e_5 \end{pmatrix}.
$$

Estimating parameters using this design matrix, one obtains

$$\text{Years} = 8.75 - 0.75 * \text{Type of Car} + \text{Residual}.$$

Both the t test for the slope coefficient and the coefficient of determination do not change from the results obtained using dummy coding. We therefore come to the same conclusion as before. What has changed are the values of the regression coefficients. The intercept parameter now estimates the overall mean of the five cars in the analysis, plus $b * X_1$, and the slope coefficient estimates how much length of usage [in years] differs for the two types of cars from this overall mean. For pickups we estimate, as before, $8.75 - 0.75 * 1 = 8$ years, and for sedans we estimate $8.75 - 0.75 * (-1) = 9.5$ years, also as before. Again, the difference between the two car types is 1.5 years.

It is often seen as problematic in this analysis that the intercept parameter is not a very reliable estimate of overall length of usage, as it depends on more sedans than pickup trucks. Imagine that we have data available from 998 sedans and only two pickup trucks. When we calculate the average length of usage for the 1000 cars, this mean would be just an estimate of the mean for sedans as the two values of the pickup trucks virtually do not influence the overall mean. In cases where group sizes differ it is often more meaningful to calculate a weighted mean that reflects group size.

This can be accomplished by selecting scores for the effect coding vector that discriminates groups such that the sum of scores is zero.

The following example analyzes the car-type data again. Using effect coding with equal sums of weights for the two types of cars we arrive at the following regression equation:

$$
\begin{pmatrix} 6 \\ 10 \\ 8 \\ 9 \\ 10 \end{pmatrix} = \begin{pmatrix} 1 & 1 \\ 1 & 1 \\ 1 & 1 \\ 1 & -1.5 \\ 1 & -1.5 \end{pmatrix} \begin{pmatrix} b_0 \\ b_1 \end{pmatrix} + \begin{pmatrix} e_1 \\ e_2 \\ e_3 \\ e_4 \\ e_5 \end{pmatrix}.
$$

Once again, the t test for the slope parameter and R^2 have not changed. The interpretation of the intercept and the slope parameter is the same as in the last model, but the intercept is now a more reliable estimate of overall length of usage of sedans and pickup trucks. The regression equation is now

$$\text{Years} = 8.6 - 0.6 * \text{Type of Car} + \text{Residual.}$$

This equation suggests that the estimated duration for pickups is $8.6 - 0.6 * 1 = 8$ years, and for sedans it is $8.6 - 0.6 * (-1.5) = 9.5$ years, which is the same as before.

What should be noted from these comparisons is that different coding schemes do change values and interpretation of regression coefficients, that is, the parameters of the model – hence the name reparameterization. However, model fit, predicted values, and the significance test for the slope parameter stay the same, regardless of which coding scheme is employed.

This section covered the situation in which researchers attempt to predict one continuous (or categorical, dichotomous) variable from one dichotomous predictor. The following section extends this topic in two ways. First, it covers predictors with more than two categories. Second, it covers multiple predictors.

4.2   More Than Two Categories

Many categorical predictors have more than two categories. Examples include car brand, religious denomination, type of cereal, type of school, topic to study, type of mistake to make, race, citizenship, ice cream flavor, swimming stroke, soccer rule, and belief system.

To be able to use categorical predictors in regression analysis one creates dummy coding or effect coding variables as was previously explained. Recall that the way in which the predictors are coded determines the way the regression coefficient is interpreted. Therefore, if the coding of the design matrix can be done such that the parameters to be estimated correspond to research hypotheses of interest, then these hypotheses tests are equivalent to the tests of the related regression coefficient. How this can be achieved is demonstrated in this section by applying effect coding of multicategory variables.

Consider a predictor with k categories. For k categories one can create up to $k-1$ independent contrasts. Contrasts compare one set of categories with another set, where each set contains one or more categories. There are several ways to create contrasts for these k categories. If the number of cases per category is the same, researchers often create orthogonal contrasts, that is, contrasts that are independent of each other. If the number of cases per category differs, researchers typically create contrasts that reflect category comparisons of interest and either accept bias in the parameter estimates or deal with multicollinearity.

Orthogonal Contrasts

To begin with an example, consider the variable Swimming Stroke with the four categories Back Stroke (BS), Butterfly (BF), Breast (BR), and Crawl (CR). Since we focus in the following paragraphs on the mechanics of constructing a design matrix we do not have to consider a particular dependent variable. The meaning of contrasts will be illustrated in the subsequent data examples. To create orthogonal contrasts one proceeds as follows

1. Select a pair of categories to contrast. This first selection is arbitrary from a technical viewpoint. The following orthogonal contrasts may not be, depending on the number of categories. In the present example, we select the category pair BS – BF. The effect coding variable identifies members of this pair by assigning value +1 to members of one pair and -1 to members of the other pair. If one group contains more categories than the other, values other than 1 and -1 can be assigned. For instance, if one group contains four categories and the other group contains two categories, the scores for each category in the first group can be +0.5, and the scores for each category in the second group can be -1. The values assigned must always sum up to zero. Cases that do not belong to either pair are assigned a 0. For the present example we thus create the contrast vector $c' = (1, -1, 0, 0)$.

2. Repeat Step 1 until either all $k - 1$ contrast vectors have been created or the list of contrasts that are of interest is exhausted, whatever comes first.[2] If the goal is to create orthogonal contrasts, it is not possible to create $k - 1$ contrasts where each comparison involves only two single-categories. One must combine categories to create orthogonal contrasts. Specifically, for even k one can create $k/2$ single-category orthogonal contrasts. For odd k one can create $(k - 1)/2$ single-category orthogonal contrasts. For the present example we can create $4/2$ single-category orthogonal contrasts, for example $c'_1 = (1, -1, 0, 0)$ and $c'_2 = (0, 0, 1, -1)$. Alternatively, we could have created $c'_1 = (1, 0, -1, 0)$ and $c'_2 = (0, 1, 0, -1)$, or $c'_1 = (1, 0, 0, -1)$ and $c'_2 = (0, 1, -1, 0)$. All contrasts after single category orthogonal contrasts combine two or more categories. (One can, however, start creating contrasts with combinations of categories and use single category contrasts only when necessary.) One complete set of $k - 1 = 3$ orthogonal contrasts for the present example is $c'_1 = (1, -1, 0, 0)$, $c'_2 = (0, 0, 1, -1)$, and

---

[2]Note that some authors recommend including contrasts even if they are not of particular substantive interest if they allow researchers to bind significant amounts of systematic variance. The price for each of these vectors is one degree of freedom. The benefit is that the residual variance will be reduced and, therefore, the contrasts of particular interest stand a better chance of capturing significant portions of variance.

$c_3' = (0.5, 0.5, -0.5, -0.5)$. It should be noted that the term orthogonality refers to the columns of the design matrix and not to the orthogonality of the c_i'. Only if the number of observations in each group is the same is the orthogonality of c_i' and c_j' (where $i \neq j$) equivalent to the orthogonality of the corresponding columns of the design matrix. Note that some authors recommend that for weighting purposes the sum of all absolute values of contrast values be a constant. Therefore, the last vector would contain four 0.5 values rather than four 1's. If the absolute values of coefficients are of no interest, the following specification of c_3' is equivalent: $c_3' = (1, 1, -1, -1)$.

3. Create coding vectors for the design matrix. Each contrast vector corresponds to one coding vector in the design matrix. Contrast vectors, c_i', are translated into design matrix coding vectors, x_i, by assigning each case the same value as its group in the contrast vector. In the present example, each case that belongs to category BS is coded 1 for the first coding vector, 0 for the second coding vector, and 0.5 for the third coding vector. Accordingly, each case that belongs to category BF is coded -1 for the first coding vector, 0 for the second coding vector, and 0.5 for the third coding vector.

Suppose a researcher investigates four athletes in each of the four swimming stroke categories. Using the above set of coding vectors, the design matrix displayed in Table 4.2 results.

Readers are invited to create alternative design matrices using the alternative first two contrast vectors listed above.

To make sure the coding vectors in the design matrix are orthogonal, one calculates the inner product of each pair of vectors. Specifically, there are $\binom{m}{2}$ pairs of coding vectors, where $m = k - 1$. If the inner product equals zero, two vectors are orthogonal. For example, the inner product, $x_1'x_3$, of the first and the last vectors is $x_1'x_3 = 1 * 0.5 + 1 * 0.5 + 1 * 0.5 + 1 * 0.5 - 1 * 0.5 - 1 * 0.5 - 1 * 0.5 - 1 * 0.5 - 0 * 0.5 - 0 * 0.5 - 0 * 0.5 - 0 * 0.5 - 0 * 0.5 - 0 * 0.5 - 0 * 0.5 - 0 * 0.5 = 0$. Readers are invited to calculate the inner products for the other two pairs of vectors, $x_1'x_2$ and $x_2'x_3$.

Table 4.2: *Design Matrix for Analysis of Four Swimming Strokes*

Case	Coding Vectors		
	x_1	x_2	x_3
1	1	0	0.5
2	1	0	0.5
3	1	0	0.5
4	1	0	0.5
5	-1	0	0.5
6	-1	0	0.5
7	-1	0	0.5
8	-1	0	0.5
9	0	1	-0.5
10	0	1	-0.5
11	0	1	-0.5
12	0	1	-0.5
13	0	-1	-0.5
14	0	-1	-0.5
15	0	-1	-0.5
16	0	-1	-0.5

Selection and Characteristics of Contrasts

In the following paragraphs we discuss a number of issues related to selection and characteristics of contrasts in regression.

1. As was indicated earlier, the selection of the first contrast is largely arbitrary. Therefore, researchers often select the substantively most interesting or most important groups of categories for the first pair.

2. If one creates orthogonal contrasts for k categories, the kth contrast is always a linear combination of the $k-1$ orthogonal ones. Consider the following example. For a three-category variable, we create the following two orthogonal contrast vectors: $c_1' = (1, -1, 0)$ and $c_2' = (0.5, 0.5, -1)$. A third vector could be $c_3' = (1, 0, -1)$. Vector c_3 is linearly dependent upon c_1' and c_2' because $c_3' = 0.5c_1' + c_2'$. Specifically, we obtain for the first value of c_3, $1 = 0.5 * 1 + 0.5$; for the second value of c_3, $0 = 0.5 * (-1) + 0.5$; and for the third

value of c_3, $-1 = 0.5 * 0 - 1$. It is important to note that the bivariate correlations among linearly dependent variables are not necessarily equal to one. The correlations among the three vectors c_1, c_2, and c_3 are $c_1'c_2 = 0.00$, $c_1'c_3 = 0.50$, and $c_2'c_3 = 0.87$. Thus, inspection of correlation matrices does not always reveal patterns of linear dependencies. Only if correlations are numerically zero, that is, $r = 0.0$, can variables be assumed to be independent.

3. The sign given to categories is arbitrary also. Reversing the sign only reverses the sign of the parameter estimated for a particular vector. Orthogonality and magnitude of parameter estimates will not be affected, and neither will results of significance testing. Signs are often selected such that those categories that are assumed to predict larger values on Y are assigned positive values in the coding vectors. In other words, the signs in coding vectors often reflect the direction of hypotheses.

4. Effect coding vectors that are created the way introduced here are identical to effect coding vectors created for analysis of variance main effects. Defining analysis of variance main effects using this type of effect coding vectors is known as the regression approach to analysis of variance (Neter et al., 1996).

5. There is no necessity to always create all possible effect coding vectors. If researchers can express the hypotheses they wish to test using fewer than $k - 1$ vectors, the remaining possible vectors do not need to be included in the design matrix.

The following data example analyzes user-perceived durability of German cars, measured in number of miles driven between repairs. Five brands of cars are considered: Volkswagen (V), Audi (A), BMW (B), Opel (O), and Mercedes (M). Five customers per brand responded to the question as to how many miles they typically drive before their car needs repair. Figure 4.2 displays the distribution of values.

Note that in this example the sample sizes in each category are equal. The figure suggests that customers perceive BMW and Mercedes cars as equally durable – more durable than the other three brands that seem to be perceived as equal. To compare these responses using regression

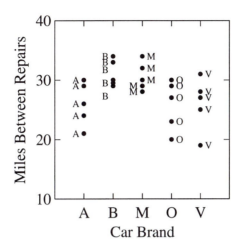

Figure 4.2: Perceived durability of cars.

analysis we have to first specify contrast vectors. For the present example we create the following orthogonal vectors:

$c_1' = (1, -1, 0, 0, 0)$,
$c_2' = (0, 0, 1, -1, 0)$,
$c_3' = (1/2, 1/2, -1/2, -1/2, 0)$, and
$c_4' = (1/4, 1/4, 1/4, 1/4, -1)$.

The first of these vectors compares user perceptions of Audis and BMWs. The second vector compares Mercedes and Opels. The third compares perceptions of Audi and BMW in one group with Mercedes and Opel in the other. The fourth contrast juxtaposes the largest German car producer, Volkswagen, and all the other brands in the sample.

After specifying contrasts one translates the contrast vectors into coding vectors. Each respondent is assigned the values of the respective contrast category. Table 4.3 displays the raw data and the resulting coding vectors. The constant vector, consisting of only ones, is omitted in the table.

Using the data from Table 4.3 we estimate parameters for the following multiple regression equation:

Miles Between Repairs $= b_0 + b_1 x_1 + b_2 x_2 + b_3 x_3 + b_4 x_4 +$ Residual.

Table 4.3: *Perceived Durability of Cars: Raw Data and Design Matrix*

Car		Effect Coding Vectors			
Brand	Miles	x_1	x_2	x_3	x_4
A	24	1	0	1/2	1/4
A	21	1	0	1/2	1/4
A	29	1	0	1/2	1/4
A	26	1	0	1/2	1/4
A	30	1	0	1/2	1/4
B	29.5	-1	0	1/2	1/4
B	29	-1	0	1/2	1/4
B	34	-1	0	1/2	1/4
B	30	-1	0	1/2	1/4
B	33	-1	0	1/2	1/4
M	28	0	1	-1/2	1/4
M	29	0	1	-1/2	1/4
M	34	0	1	-1/2	1/4
M	30	0	1	-1/2	1/4
M	32	0	1	-1/2	1/4
O	20	0	-1	-1/2	1/4
O	23	0	-1	-1/2	1/4
O	29	0	-1	-1/2	1/4
O	30	0	-1	-1/2	1/4
O	27	0	-1	-1/2	1/4
V	19	0	0	0	-1
V	25	0	0	0	-1
V	27	0	0	0	-1
V	28	0	0	0	-1
V	31	0	0	0	-1

Table 4.4: *Regression Analysis of Perceived Car Durability*

Predictor	Parameter	Std. Error	t value	p value, 2-tailed
Constant	27.90	0.71	39.60	< 0.01
x_1	-2.55	1.11	-2.29	0.03
x_2	2.40	1.11	2.15	0.04
x_3	0.35	1.58	0.22	0.83
x_4	1.90	1.41	1.35	0.19

Results of this analysis appear in Table 4.4. The portion of criterion variance accounted for is $R^2 = 0.37$. Table 4.4 suggests that two parameters are unequal to zero. Specifically, parameters for x_1 and x_2 are statistically significant. Since each regression parameter corresponds to the hypothesis associated with the corresponding contrast vector, we can conclude that BMWs are perceived as more durable than Audis, and that Mercedes are perceived as more durable than Opels.

Readers are invited to

1. recalculate this analysis including only the first two coding vectors and to calculate whether omitting the last three vectors leads to a statistically significant loss in explained variance;

2. recalculate this analysis using different sets of coding vectors; for instance, let the first two contrast vectors be $c_1 = (0, 1, -1, 0, 0)$ and $c_2 = (1, 0, 0, 0, -1)$. (Hint: the remaining three contrast vectors can be constructed in a fashion analogous to Table 4.3;)

3. recalculate this analysis using a program for ANOVA and compare the portions of variance accounted for.

4.3 Multiple Categorical Predictors

In general, one can treat multiple categorical predictors just as single categorical predictors. Each categorical predictor is transformed into effect coding variables. Specifically, let the jth predictor have k_j categories,

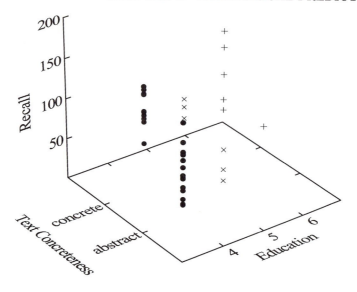

Figure 4.3: Scatterplot of the predictors, Educational Level and Text Concreteness, and the criterion, Recall.

with $k_j > 1$. Then, the maximum number of independent coding variables for this predictor is $k_j - 1$.

Interactions among predictors are rarely considered in regression analysis. However, interactions among coding variables can easily be taken into account, in particular when the number of cases for each category is the same, that is, in orthogonal designs. Chapter 10 covers interactions in regression in detail.

In the following paragraphs we present a data example that involves two categorical predictors. The data describe Recall Rates, REC, of $n = 37$ participants of a memory experiment. From the number of participants it should already be clear that this time the numbers within each group cannot be the same. The demographic variables describing the participants included the variable Education (4, high school; 5, baccalaureate; 6, masters or equivalent). Subjects read texts (Text Groups, TG) that were either concrete ($TG = 1$) or abstract ($TG = 2$). Figure 4.3 presents the scatterplot of the predictors, TG and $EDUC$, and the criterion, REC.

Table 4.5: *Regression of Recall on Education and Text Group*

Variable	Coefficient	Std. Error	t Value	p Value
Intercept	147.00	14.42	10.19	< 0.01
$EDUC_1$	7.16	6.26	1.15	0.26
$EDUC_2$	12.98	8.86	1.46	0.15
TG	−45.12	9.43	−4.78	< 0.01

The plot suggests that concrete texts are recalled better than abstract texts. In contrast, Education does not seem to have any effect on Recall. To test the predictive power of $EDUC$ and TG we first create contrast and coding vectors for $EDUC$. There is no need to create contrast or coding vectors for TG, because it only has two categories. Thus, the only possible contrast compares these two categories, and the contrast vector for TG is $\mathbf{c'_{TG}} = (1, -1)$. When the number of respondents in the two categories of TG is unequal, the coding of TG is arbitrary and has no effect on results (except for scaling of the parameter estimate). If, however, this number is equal, the above contrast coding for TG will yield a contrast vector that is orthogonal to the other vectors in the design matrix, $\mathbf{X}$.

For the three categories of $EDUC$, we create the following two contrast vectors: $\mathbf{c'_1} = (1, -1, 0)$ and $\mathbf{c'_2} = (-0.5, -0.5, 1)$. These two contrast vectors are inserted into the design matrix, along with the codes for TG. The regression equation that we now estimate is

$$\text{Recall} = b_0 + b_1 * \text{EDUC}_1 + b_2 * \text{EDUC}_2 + b_3 * \text{TG} + \text{Residual},$$

where $EDUC_1$ is the coding vector from $\mathbf{c_1}$ and $EDUC_2$ is the coding vector from $\mathbf{c_2}$. Parameter estimates and their standard errors, t values, and tail probabilities (two-tailed) appear in Table 4.5.

The multiple $R^2 = 0.482$ suggests that we can explain 48.2% of the variation of Recall Rates from Education and Text Group. Table 4.5 indicates that Text Group is the only variable that has a statistically significant effect on Recall. The two $EDUC$ variables do not have statistically significant effects. This result confirms the impression we had when inspecting Figure 4.3.

In the following paragraphs we ask whether the coding vectors in $\mathbf{X}$

are orthogonal or are correlated. If they are correlated they could possibly suffer from multicollinearity. Vector correlations are possible because the predictor categories did not occur at equal frequencies. Specifically, we obtain the following univariate frequency distributions for the predictors:

$$EDUC_1 \quad = \quad (\text{-1: 6 times; 0: 6 times; 1: 25 times}),$$
$$EDUC_2 \quad = \quad (\text{-0.5: 31 times; 1: 6 times}),$$
$$TG \quad = \quad (1 : 19 \text{ times}; 2 : 18 \text{ times}).$$

From the uneven distribution of the Education categories one cannot expect predictors to be orthogonal. Indeed, correlations are nonzero. Table 4.6 displays the correlations among the predictors, $EDUC_1$, $EDUC_2$, and TG, and the criterion, REC.

It is obvious from these correlations that the strategy that we use to create contrast vectors leads to orthogonal coding vectors only if variable categories appear at equal frequencies. The magnitude of correlations suggests that there might be multicollinearity problems.

Table 4.6: *Spearman Correlations among the Predictors, $EDUC_1$, $EDUC_2$, and TG, and the Criterion, Recall*

	$EDUC_1$	$EDUC_2$	TG
$EDUC_2$	-0.47		
TG	0.17	-0.28	
REC	-0.01	0.25	-0.69

Chapter 5

OUTLIER ANALYSIS

While in many instances researchers do not consider fitting alternative functions to empirical data, outlier analysis is part of the standard arsenal of data analysis. As will be illustrated in this chapter, outliers can be of two types. One is the distance outlier type, that is, a data point extremely far from the sample mean of the dependent variable. The second is the leverage outlier type. This type is constituted by data points with undue leverage on the regression slope. The following sections first introduce leverage outliers and then distance outliers.

5.1 Leverage Outliers

Leverage outliers are defined as data points that exert undue leverage on the regression slope. Leverage outliers' characteristics are expressed in terms of the predictor variables. In general, a data point's leverage increases with its distance from the average of the predictor. More specifically (for more details on the following sections see Neter et al., 1996), consider the *hat matrix*, $\mathbf{H}$,

$$\mathbf{H} = \mathbf{X}(\mathbf{X'X})^{-1}\mathbf{X'},$$

where $\mathbf{X}$ is the design matrix. $\mathbf{X}$ has dimensions $n \times p$, where p is the number of predictors, including the term for the constant, that is, the intercept. Therefore, $\mathbf{H}$ has dimensions $n \times n$. It can be shown that the estimated values, $\hat{y}_i$, can be cast in terms of the hat matrix and the criterion variable, Y,

$$\hat{\mathbf{y}} = \mathbf{H}\mathbf{y}, \tag{5.1}$$

and that, accordingly, the residuals can be expressed as

$$\mathbf{e} = \mathbf{y} - \mathbf{H}\mathbf{y}.$$

From (5.1) it can be seen that matrix $\mathbf{H}$ puts the hat on $\mathbf{y}$. The ith element of the main diagonal of the hat matrix is

$$h_{ii} = \mathbf{x}_\mathbf{i}'(\mathbf{X}'\mathbf{X})^{-1}\mathbf{x}_\mathbf{i}, \tag{5.2}$$

where $\mathbf{x}_\mathbf{i}$ is the ith row in $\mathbf{X}$, that is, the row for case i. The element h_{ii} has the following properties:

1. It has range $0 \leq h_{ii} \leq 1$.

2. The sum of the h_{ii} is p:

$$\sum_{i=1}^{n} h_{ii} = p, \quad \text{for} \quad i = 1, \ldots, n;$$

3. h_{ii} indicates how far the x value of case i lies from the mean of X.

4. h_{ii} is known as the leverage of case i. Large values of h_{ii} indicate that the x value of case i lies far from the arithmetic mean of X; if h_{ii} is large, case i has a large influence on determining the estimated value $\hat{y}_i$.

5. The variances of residual e_i and h_{ii} are related to each other as follows:

$$\sigma^2(e_i) = \sigma^2(1 - h_{ii}).$$

From (5.2) one can see that the variance of a residual depends on the value of the predictor for which the residual was calculated. Specifically, $\sigma^2(e_i)$ decreases as h_{ii} increases. Thus, we have zero variance for e_i if

$h_{ii} = 1$. In other words, when h_{ii} increases, the difference between the observed and the estimated y values decreases, and thus the observed y value will lie closer to the regression line. The observed y value lies exactly on the regression line if $h_{ii} = 1$. Because of the dependence of the variance of a residual upon the predictor value one often uses the studentized residual, which is defined as $e_i/\sigma^2(e_i)$, for comparisons.

In empirical data analysis the following rules of thumb are often used when evaluating the magnitude of leverage values:

1. h_{ii} is large if it is more than twice as large as the average leverage value. The average leverage value is

$$\bar{h} = \frac{1}{n} \sum_{i=1}^{n} h_{ii} = \frac{p}{n}.$$

2. $h_{ii} > \frac{1}{2}$ indicates very high leverage; $0.2 \leq h_{ii} \leq 0.5$ suggests moderate leverage.

3. There is a gap between the leverage for the majority of cases and a small number of large leverage values.

For the following example we use data from the Finkelstein et al. (1994) study. Specifically, we use the scores in verbal aggression (*VA85*) and physical aggression against peers (*PAAP85*), collected in 1985. The sample consists of $n = 77$ boys and girls. We ask whether verbal aggression allows one to predict physical aggression. The following parameters are estimated for this regression problem:

PAAP85 = 9.51 + 0.46 * VA85 + Residual.

The slope parameter is significantly greater than zero ($t = 4.07; p < 0.01$). The scatterplot of *VA85* and *PAAP85* appears in Figure 5.1.

The figure shows that most of the data points nestle nicely around the regression line. It also suggests that applying a linear regression line matches data characteristics. However, there are two data points that are abnormal. One of these points describes a boy with slightly above average verbal aggression but a very large number of physical aggression acts against peers. While an outlier on the dependent variable's scale, this

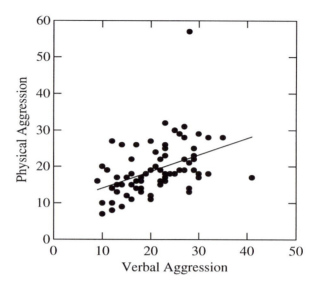

Figure 5.1: Regression on physical aggression on verbal aggression in Adolescents.

data point has no strong influence on the regression slope. In contrast, the data point in the lower right corner of the figure does have above average leverage. It is an outlier on the predictor variable's scale, and thus a leverage point with a leverage value of $h = 0.126$.

We now analyze the leverage values for the present sample. The mean of leverage values is $= 2/77 = 0.026$ with a minimum of $h = 0.013$ and a maximum of $h = 0.126$. The median is $md = 0.021$. Clearly, the leverage point meets the first of the above criteria by being more than twice as large as the average leverage. In addition, it meets the third criterion in that there is a gap between the crowd of leverage points and this individual point. This is illustrated by the histogram in Figure 5.2.

Figure 5.2 shows that the majority of the leverage values is grouped around the median. The leverage value of this boy appears only after a large gap.

For a more in-depth analysis of the effects of this leverage point we reestimate the regression parameters after excluding it. The resulting

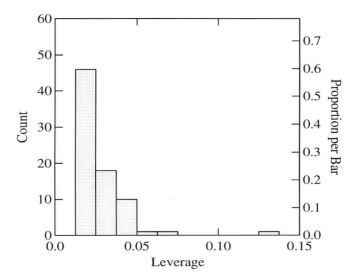

Figure 5.2: Bar graph of leverage scores of data points in Figure 5.1.

regression equation is

$$PAAP85 = 8.113 + 0.53 * VA85 + Residual.$$

The slope parameter is significantly greater than zero ($t = 4.52; p < 0.01$). A comparison with the original regression parameter estimates suggests that the leverage point did indeed "pull the slope down." Figure 5.3 shows the new regression slope for the data without the leverage point.

Inspection of leverage values for this analysis does not reveal any additional or new leverage point. Only the outlier, high up in the number of aggressive acts, is still there.

Distance Outliers

Leverage outliers are cases with extreme values on the predictor variable(s). Distance outliers are cases with extreme values on the criterion variable. They can be identified using the studentized residuals, that is, residuals transformed to be distributed as a t statistic. Specifically, for

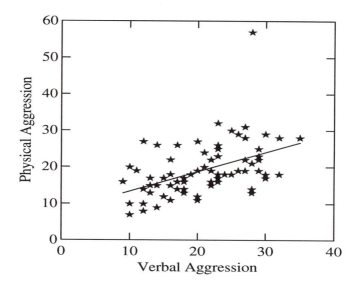

Figure 5.3: Regression of physical aggression on verbal aggression after removal of leverage point.

each case, the studentized residual, d_i^*, is given by

$$d_i^* = e_i \sqrt{\frac{n - p - 1}{(\sum_{i=1}^{n} e_i^2)(1 - h_{ii}) - e_i^2}}, \tag{5.3}$$

where p is the number of parameters estimated by the regression model, including the intercept parameter. Studentized residuals have the following characteristics:

1. Each value d_i^* is distributed as a t statistic with $n - p - 1$ degrees of freedom.

2. The d_i^* are not independent of each other.

3. To calculate the d_i^* one only needs the residuals and the h_{ii} values.

To determine whether a given value, d_i^*, is an outlier on the criterion, one (1) calculates the critical t value for α and the $n - p - 1$ degrees of freedom and (2) compares the d_i^* with this value, $t_{\alpha, df}$. If $d_i^* > t_{\alpha, df}$, case i is a distance outlier.

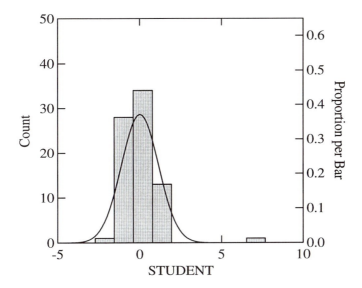

Figure 5.4: Density histogram of studentized residuals for data in Figure 5.1

For a numerical example we use the verbal – physical aggression data from Figure 5.3. The critical t value for $df = 76-2-1 = 73$ and $\alpha = 0.05$ is $t_{0.05,73} = 1.666$. The density histogram for the 76 cases appears in Figure 5.4.

The histogram shows that there is one case that is far out on the t distribution scale. This is the case with the very large number of physical aggressive acts in Figure 5.3. This boy's studentized residual is $d_i^* = 6.77$; this value is clearly greater than the critical t. However, the figure also suggests that there are more cases with d_i^* values greater than the critical one. We find two such cases. The first (order is of no importance) has values $VA85 = 12$, $PAAP85 = 27$, and $d_i^* = 1.85$. This is the case in the upper left of the data cloud in Figure 5.3. The second has values $VA85 = 23$, $PAAP85 = 32$, and $d_i^* = 1.85$. This is the case with the second highest $PAAP85$ score in the distribution.

These two data points suggest that it is both the absolute magnitudes of scores on Y and the difference between the observed and the estimated scores that determine whether or not a case is a distance outlier in terms of Y.

Cook's Distance, D_i

Cook's distance is a measure of the magnitude of the influence that case i had on all of the parameter estimates in a regression analysis. The measure is defined as

$$D_i = \frac{e_i^2 h_{ii}}{p \frac{1}{n} (\sum_{i=1}^n e_i^2)(1 - h_{ii})^2},$$ (5.4)

where p is the number of parameter estimates, including the intercept parameter.

Cook's distances, D_i, have the following characteristics:

1. Although D_i is not distributed as an F statistic, it usually is evaluated in regard to F_α with p and $n - p$ degrees of freedom.

2. The following rules of thumb apply:

   - When $p(D_i) < 0.10$, case i is said to have little influence on the magnitude of parameter estimates;

   - When $p(D_i) > 0.50$, case i is said to have considerable influence on the fit of the regression model

3. As the d_{ii}, the D_i can be calculated from the residuals and the h_{ii} values.

Instead of giving an application example of Cook's distance, we illustrate the relationship between Cook's distance and the studentized residual. A comparison of Formulas (5.3) and (5.4) suggests that, while d_{ii} and D_i use the same information, they process it in different ways. Specifically, the relationship between the two measures does not seem to be linear. This is illustrated in Figure 5.5.

Figure 5.5 displays the scatterplot of the studentized residuals and the Cook distances for the data in Figure 5.1. In addition to the data points, the figure shows the quadratic regression line for the regression of the Cook distances onto the studentized residuals. Obviously, the fit is very good, and both measures identify the same cases as extremes.

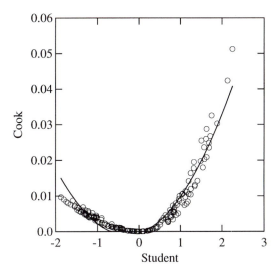

Figure 5.5: Scatterplot and quadratic curve for regression of the Cook statistic on studentized residuals.

Around the value of $d_{ii} = 0$ the data are denser and follow a quadratic pattern. The outlier is identified by both measures. Thus, we conclude that, while d_{ii} and D_i will not always suggest the same conclusions, their relationship is strong and they tend to agree in outlier identification.

5.2 Remedial Measures

Thus far, this chapter has presented methods for identifying specific problems with the data - model fit when employing a particular regression model. It is most important to note that any of these problems is defined only with respect to the data estimation methods and statistical evaluation methods employed. Using other methods, the problems may not surface or even exist. For example, the problem with the residuals not being much different than the raw scores in the two panels of Figure 6.1 could have been avoided by using curvilinear regression.

Nevertheless, the question is what one can do to overcome or remedy problems with the data - model fit. Obviously, problems with this fit can lead to misestimations of parameters and their standard errors and thus to misrepresentations of variable relationships. There is no common cure for lack of data - model fit. Each problem must be approached by specific remedial measures. The following sections present remedial measures for specific problems with the model - data fit.

5.2.1   Scale Transformations

When data show a pattern that is nonlinear, as, for example, in Figures 6.1 and 8.1, researchers may wish to consider either of the following measures:

1. Application of curvilinear regression;

2. Data transformation.

Solution 1, the application of curvilinear regression, is typically approached either from an exploratory perspective or from a theory-guided perspective. The exploratory approach is where researchers try out a number of types of curves and select the one that yields the smallest sum of squared residuals, that is, the smallest value for the least squares criterion (see Section 2.2.2). While appealing in many respects, this approach is seen by many as too close to data fitting, that is, as an approach where functions are fit to data with no reference to theory.

Theory-guided approaches to curvilinear regression start from a type of function for a shape that a curve can possibly take. Then, the data are fit to a selection of such functions. Only functions with interpretable parameters are selected, and each of these functions corresponds to a particular interpretation of substantive theory.

The second measure, data transformation, is a widely used and widely discussed tool. It can serve to deal not only with problems of nonlinearity, but also with problems of unequal error variances, skewness of the distribution of error terms, and nonnormality. The best known group of procedures to deal with the first three of these problems are known as Box-Cox transformations (Box & Cox, 1964; Cox, 1958; Neter et al., 1996; Ryan, 1997). Box-Cox transformations use power transformations of Y into Y' of the kind

$$Y' = Y^\lambda,$$

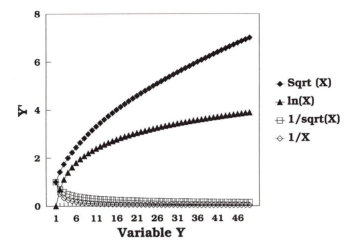

Figure 5.6: Sample transformations of Y.

where λ is a parameter that is determined specifically for a given data set. Examples of λ and the resulting transformed variables Y' appear in Table 5.1 (cf. Ryan, 1997).

The effects of the last four of these nonlinear transformations[1] on a variable Y that ranges from 1 to 50 (in steps of 1) are illustrated in Figure 5.6.

Table 5.1: *Examples of Power Transformations of the Criterion Variable, Y*

Value of Parameterλ	Transformed Variable Y'
2	$Y' = Y^2$
1/2	$Y' = \sqrt{Y}$
0	$Y' = \log Y$ (by definition)
-1/2	$Y' = 1/\sqrt{Y}$
-1	$Y' = 1/Y$

---

[1]It should be noted that these transformations do not vary monotonically depending on λ when $Y' = \log Y$ for $\lambda = 0$. Therefore, Box and Cox (1964) specified that $Y' = (Y^\lambda - 1)/\lambda$ if $\lambda \neq 0$, and $Y' = \log Y$ if $\lambda = 0$.

Explanations of these transformations follow below. A typical application of Box-Cox transformations aims at minimizing the residual sum of squares. There is computer software that performs the search for the optimal transformation.

However, in many instances, researchers do not need to search. Rather, there is a data problem that can be solved by a specific transformation. In addition, some transformations help solve more than just one data problem. The following paragraphs describe examples.

The Square Root Transformation

Both across the categories of categorical predictors and for continuous predictors it can occur that means and standard deviations of the criterion variable are functionally related. This is the case for Poisson processes, that is, rare events assessed by counting, given certain assumptions like independence of events for different time intervals. The standard square root transformation is

$$Y' = \sqrt{Y}.$$

When there are measures $0 \leq Y < 10$, the following transformation seems more suitable:

$$Y' = \sqrt{Y + 1/2}.$$

Square root transformations stabilize, that is, render more homogeneous, variances and, in addition, normalize distributions.

The Logarithmic Transformation

When, for a continuous predictor or across the categories of a categorical predictor, standard deviations are proportional to means, logarithmic transformations may be considered. Specifically, one transforms Y into Y' by

$$Y' = \log Y,$$

where log is the natural logarithm, that is, the logarithm with base $e = 2.71828\ldots$. When there are values $y_i = 0$ one adds to all scores a constant

Δ, where Δ is a small constant that is often set to 1 or 0.5. In other words, one uses the transformation

$$Y' = \log(Y + \Delta)$$

instead of a simple logarithmic transformation.

The logarithmic transformation is among the most frequently employed transformations. Some of the scales we use everyday, for example, the phone scale for acoustic intensity, are logarithmic scales. In psychology, GSR (galvanic skin resistance) scales are typically logarithmic also.

Trigonometric Transformation

Chiefly to reduce instability of variances when the dependent measures are proportions, one employs trigonometric transformations. The best known of these is the inverse sine transformation

$$Y' = \arcsin \sqrt{Y}. \tag{5.5}$$

Often researchers multiply the term on the right-hand side of (5.5) by 2 and add a constant to Y when there are values $Y = 0$.

Reciprocal Transformation

When the squares of means are proportional to some unit of Y, one may consider a reciprocal transformation such as

$$Y' = 1/Y \quad \text{or} \quad Y' = \frac{1}{Y+1},$$

where the second of these equations is appropriate when there are values $y = 0$. The reciprocal transformation is often employed when the dependent measure is time, for example, response times or problem solving times.

Many more transformations have been discussed, in particular, transformations to correct lack of normality (cf. Kaskey, Koleman, Krishnaiah, & Steinberg, 1980; Ryan, 1997). It is most important to realize that nonlinear transformations can change many variable characteristics. Specifically, nonlinear transformations not only affect mean and standard deviations, but also the form of distributions (skewness and kurtosis) and additivity of effects. In addition, the power of tests on transformed variables may be affected (Games, 1983, 1984; Levine & Dunlap, 1982, 1983). Therefore, routine application of transformations can have side effects beyond the desired cure of data problems. Researchers are advised to make sure data have the desired characteristics after transformation without losing other, important properties.

5.2.2   Weighted Least Squares

This section describes one of the most efficient approaches to dealing with unequal variances of the error terms: Weighted Least Squares (WLS) (for the relationship of WLS to the above transformations see, for instance, Dobson (1990, Chapter 8.7)). In Section 2.2.2 the solution for the ordinary least squares minimization problem, that is,

$$(\mathbf{y} - \mathbf{Xb})'(\mathbf{y} - \mathbf{Xb}) \longrightarrow \min,$$

was given as

$$\mathbf{b} = (\mathbf{X}'\mathbf{X})^{-1}\mathbf{X}'\mathbf{y},$$

where the prime indicates transposition of a matrix. The main characteristic of this solution is that each case is given the same weight. Consider the weight matrix, $\mathbf{W}$,

$$\mathbf{W} = \begin{pmatrix} w_{11} & 0 & \ldots & 0 \\ 0 & w_{22} & \ldots & 0 \\ \vdots & \vdots & \ddots & \vdots \\ 0 & 0 & \ldots & w_{nn} \end{pmatrix},$$

that is, a matrix with a weight for each case. Then, the OLS solution can be equivalently rewritten as

$$\mathbf{b} = (\mathbf{X}'\mathbf{W}\mathbf{X})^{-1}\mathbf{X}'\mathbf{W}\mathbf{y}, \tag{5.6}$$

where $\mathbf{W} = \mathbf{I}$, the identity matrix. If $\mathbf{W}$ is a diagonal matrix with unequal diagonal elements, (5.6) is the WLS solution for the least squares problem.

Selection of Weights

In most instances, weights cannot be straightforwardly derived from theory or earlier results. Therefore, researchers typically select weights such that they address their particular data problem. For example, if the heteroscedasticity problem, that is, the problem with unequal error variances, is such that the variance of residuals increases with X, one often finds the recommendation to consider the following weight:

$$w_{ii} = \frac{1}{x_i}.$$

Accordingly, when the variance of residuals decreases with X, one can find the recommendation to consider the weight

$$w_{ii} = \frac{1}{x_{max} - x_i}, \tag{5.7}$$

where x_{max} is the largest value that X can possibly assume.

The following example illustrates the effects one can expect from estimating regression parameters using (5.7) and WLS. We analyze a data set with error variance depending on X. For Figure 5.7 we analyzed data that describe pubertal developmental status, operationalized by the Tanner score, *T83*, and the number of aggressive acts against peers, *PAAP83*, in a sample of $n = 106$ boys and girls in 1983, all in early puberty. The two panels of Figure 5.7 show that there is at least one outlier in this data set. In addition, the right panel suggests that the size of residuals depends heavily on the predictor, *T83*. In the following paragraphs we reestimate regression parameters for these data using WLS. Before we do this, we recapitulate the OLS regression equation obtained for these data:

$$\text{PAAP83} = 21.86 - 0.13 * \text{T83} + \text{Residual}.$$

The slope parameter for this analysis had not reached statistical significance ($t = -0.32; p = 0.747$).

The right panel of Figure 5.7 suggests that the error variance decreases as the values of *T83* increase. Therefore, we define as the weight variable $W = 1/(13 - \text{T83})$, with 13 being the largest Tanner score in this sample and age group. Using this weight variable we perform WLS regression as

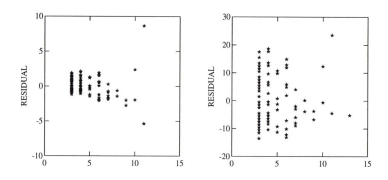

Figure 5.7: Residual plots for OLS (left) and WLS (right) analyses.

follows:

1. We multiply *PAAP83* with W to obtain

$$\text{PAAP83}_W = \text{PAAP83} * W$$

2. We multiply *TA83* with W to obtain

$$\text{TA83}_W = \text{TA83} * W$$

3. We estimate the regression parameters.

For the present example we calculate

$$\text{PAAP83} = 18.04 * W + 0.69 * \text{T83} * \text{Residual}.$$

The F test for the predictor, $T83_W$, now suggests that the slope parameter is greater than zero. Specifically, we obtain $F_{1,103} = 2.19$ and $p = 0.03$. Table 5.2 presents the parameter estimates for both solutions.

The two standard errors for the predictor parameters show the benefit from WLS most dramatically. The WLS standard error is both in absolute terms and relative to the magnitude of the parameter estimate smaller than the OLS standard error. As a result, the relationship between physical pubertal development and number of aggressive acts against peers,

Table 5.2: *OLS and WLS Parameter Estimates for Aggression Data*

Parameter	OLS Estimate	OLS Std. Error	WLS Estimate	WLS Std. Error
Constant	21.855	1.946		
Weight			18.040	2.170
$T83_W$	−0.127	0.392	0.692	0.315

inconspicuous from OLS analysis, is now statistically significant.

Figure 5.7 contains the residual plots for the OLS analyses and the present WLS solution.

The first panel of Figure 5.7 shows the residual plot for the OLS solution. It suggests that the variance of the residuals decreases as the predictor values, *T83*, increase. The right panel of Figure 5.7 shows the effects of the weighting for the WLS solution. The effect is that the error variance, that is, the variation of residuals around zero, is more even for the WLS solution than for the OLS solution. The data points are coded according to the value of the original predictor variable, *T83*. As a result, one can make out that the weighting did not change the rank ordering of data points. It only changed the scale units and reduced the error variance.

Caveats

In a fashion similar to the caveats given concerning variable transformations, caveats concerning WLS seem in order. The first concerns recommendations that are not grounded in substantive theory. Whenever weights for WLS are estimated from the data, WLS loses its desirable optimality characteristics (even though it may still be better than OLS). Relatively unproblematic may be the options to use the scatterplot of residuals with Y for weight estimation or to use the residuals for weight estimation.

As was obvious from the present example, WLS can, sometimes, be used to change the significance of results. However, WLS is not an all-encompassing cure for lack of statistical significance. The specification of weights can be as arbitrary as the selection of a transformation procedure. Therefore, researchers are well advised to switch from OLS to WLS only if either there is *a priori* knowledge of weights, e.g., in the form of a variance–covariance matrix of estimated regression coefficients, or a derivation of weights is performed on substantive or theoretical grounds. Researchers have to resist abusing such tools as variable transformation and weighted least squares for arbitrary manipulation of data.

Chapter 6

RESIDUAL ANALYSIS

Using methods of residual analysis one can determine whether

1. the function type employed to describe data reflects the relationships present in the data; for example, if researchers chose a straight line for data description, curved relationships can be captured only in part, if at all;

2. there are cases that contradict[1] this type of relationship, that is, whether there are outliers; and

3. there are anomalies in the data; examples of such anomalies include standard deviations that vary with some predictor (heteroscedasticity).

This chapter presents analysis of residuals, which can be defined as

$$e_i = y_i - \hat{y}_i,$$

where e_i is the residual for case i, y_i is the value observed for case i, and $\hat{y}_i$ is the expected value for case i, estimated from the regression equation.

Before providing a more formal introduction to residual analysis we show the benefits of this technique by providing graphical examples of cases (1), (2), and (3).

---

[1]The terms "contradicting" and "outliers" are used here in a very broad sense. Later in this chapter we introduce more specific definitions of these terms.

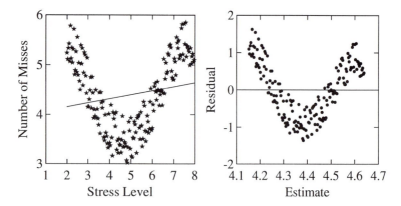

Figure 6.1: Raw data and residual plot for straight line regression of a V-shaped relationship.

6.1 Illustrations of Residual Analysis

First, we illustrate how far a chosen regression function can be from validly describing data. We select linear regression as the sample regression function. Similar examples can be constructed for curvilinear regression. Figure 6.1 gives an example (two panels). The example describes the number of misses made by professional dart players under a range of stress intensities. Stress ranged from very low as in a pub situation, to intermediate as in a local competition, to very high as in the world championships. There was a total of eight stress levels, with 1 indicating the lowest level. Levels 2 through 8 were realized in the experiment. The dependent variable was the number of misses, averaged over teams of three players in a total of $n = 180$ games.

The left-hand panel of Figure 6.1 displays the scatterplot of the misses in $n = 180$ games. The distribution of averaged misses suggests a V-shaped function of misses, depending on stress level. In addition, the curve suggests that, on average, higher stress causes more misses than lower stress. This is also indicated by the regression line, which has a positive slope. The regression function for the relationship between Number of Misses M and Stress Level S is

$$M = 3.99 + 0.08 * S + \text{Residual}.$$

The F value for the regression slope is $F = 6.66$. This value is, for $df_1 = 1$, and $df_2 = 178$, statistically significant, $p = 0.011$. The squared multiple R is smallish. It is $R^2 = 0.036$.

In standard application of regression analysis researchers may be tempted to content themselves with this result. They might conclude that there is a statistically significant linear relationship between Number of Misses and Stress Level such that increases in stress cause increases in the number of misses. However, an inspection of the residuals suggests that the linear regression model failed to capture the most important aspect of the relationship between Stress Level and Number of Misses, that is, the curvilinear aspect.

The right-hand panel of Figure 6.1 displays the residual plot (predictor x residuals) for the above regression equation. It plots Estimate against Size of Residuals. The comparison of the two panels in Figure 6.1 reveals two important characteristics of the present example:

1. The curvilinear characteristics of the raw data and the residuals are exactly the same. This does not come as a surprise because the regression model employed in the left panel is not able to capture more than the linear part of the variable relationship.

2. The increase in misses that comes with an increase in stress does not appear in the residual plot. The reason for this is that the linear regression model captured this part of the variable relationship.

Thus, we can conclude that when a regression model captures all of the systematic part of a variable relationship the residuals will not show any systematic variation. In other words, when a regression model captures all of the systematic part of a variable relationship, the residuals will vary completely at random.

This characteristic is illustrated in Figure 6.2. The left-hand panel of this figure displays two random variates that correlate to $r = 0.87$. The joint distribution was created as follows. A first variable, *NRAN1*, was created using a standard normal random number generator for $n = 100$. A second variable, *NRAN2*, was created the same way. A third variable, *NRAN3*, was created using the following formula:

$$\text{NRAN3} = \text{NRAN1} + 0.6 * \text{NRAN2}.$$

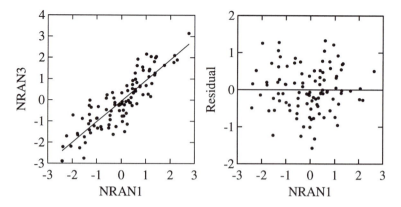

Figure 6.2: Residual plot for linear relationship.

The left-hand panel of Figure 6.2 displays the joint frequency distribution of *NRAN1* and *NRAN3*.

Because of the random character of these two variables and because of the built-in linear relationship between them, we expect no systematic variation in the residuals. This is illustrated in the right-hand panel of Figure 6.2. This panel displays a bivariate normal distribution. It is as perfect as a random number generator can create for the (relatively small) sample size of $n = 100$.

In the second example we illustrate outliers. From the many definitions of outliers, we use here distance outliers. These are data points located unusually far from the mean of the dependent measure. Among the many reasons why there are outliers, the following two are most often discussed:

1. *Measurement error.* This is the most often considered reason for the existence of outliers. The measurement instrument may have indicated wrong values or may have been misread; data typists may have hit the wrong key; coders may have miscoded a response; or respondents may have crossed the wrong answer. If the number found in the data is theoretically possible (as, for instance, an IQ of 210), it may be hard, if not impossible, to identify a value as a mistake. If, however, a value lies beyond the limits of the scale used (as, for instance, the value 9 on a rating scale with a range from 1

to 5), it can be detected relatively easily.

2. *Presence of unique processes.* If an outlier displays an extreme value that, in theory, is possible, this value is often explained as caused by unique processes. Examples of such processes include luck, extreme intelligence, cheating, and pathological processes. If any of these or similar phenomena are considered, researchers often feel the temptation to exclude outliers from further analysis. The reason given for excluding cases is that they belong to some other population than the one under study. Statistical analysis of outliers can provide researchers with information about how (un)likely a given value is, given particular population characteristics.

The following data example illustrates the presence and effects of distance outliers. Finkelstein et al. (1994) analyzed the development of aggressive behavior during adolescence in a sample of $n = 106$ adolescent boys and girls with a 1983 average age of 12 years. The authors estimated a Tanner score and an Aggression score for each adolescent. The Tanner score, T, is a measure of physical development. The higher the Tanner score, the more advanced is a child's physical development. The Aggression score measured frequency of Physical Aggression Against Peers, *PAAP83*. The higher the *PAAP83* score, the more frequent are an adolescent's aggressive acts against peers. In their analyses the authors attempted to predict aggression from physical pubertal development.

The left-hand panel of Figure 6.3 displays the scatterplot of Tanner scores and *PAAP83* scores in the sample of 106 adolescents. The plot displays the data points and the regression line. The regression function is

$$PAAP83 = 21.86 - 0.13 * T + \text{Residual}.$$

The t value[2] for the slope coefficient is $t = -0.324$ and has a tail probability of $p = 0.747$. Thus, physical development does not allow researchers to predict frequency of physical aggression against peers.

In the left-hand panel of Figure 6.3 stars below the regression line

---

[2]It should be emphasized that, in the present context, the t test and the F test for regression coefficients are equivalent. Selection of test is arbitrary.

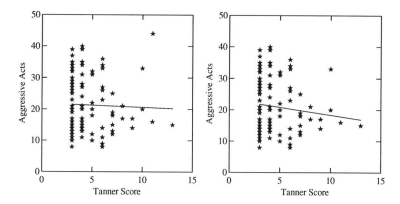

Figure 6.3: Leverage outlier and linear relationship.

indicate aggression values that are smaller than estimated from the regression function (bottom half of the scatterplot). Stars above the regression line indicate aggression values greater than estimated from the regression function (top half of the scatterplot). The largest residual was calculated for a boy with a *PAAP83* value of 44 (highest in the sample) and a Tanner score of 11 (tied for second highest in the sample). From the regression function, this boy was expected to have an aggression score of PAAP83 = 20.46. Thus, the residual is $e = 23.54$. The star for this boy appears in the upper right corner of the plot.

Assuming this boy may not only have an extremely large *PAAP83* score but also may have leveraged the slope of the regression line to be less steeply decreasing (see leverage outliers in Section 5.1), we reestimated the regression parameters under exclusion of this boy. The resulting function is

$$PAAP83 = 23.26 - 0.49 * T + \text{Residual.}$$

This function suggests a steeper decrease of aggression frequency with increasing Tanner score. Yet, it is still not statistically significant ($t = -1.25; p = 0.216$).

The right panel of Figure 6.3 displays the data points and the regression line for the reduced data set. The data points in both panels of Figure 6.3 are presented under the assumption that each point carries the

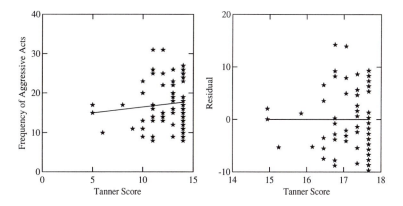

Figure 6.4: Heteroscedasticity in a linear relationship.

same weight.

The following, third data example illustrates the use of residual analysis for detection of data anomalies. One data characteristic that tends to inflate standard errors of parameter estimates is that error variances are unequal (heteroscedasticity). An example of unequal error variances is that the error variance depends on the predictor.

The following example shows error variances that increase monotonically with the level of the predictor. In this example, the regression slope is positive and the variance of the residual term increases with the expected value of the response variable. Indeed, this a very common feature of many empirical data sets. Therefore, one should always check one's data for presence of heteroscedasticity.

The data describe the $n = 69$ boys and girls available for a third examination of Tanner scores (T) and Aggression in 1987 ($PAAP87$) from the Finkelstein et al. (1994) study. The left-hand panel of Figure 6.4 presents the scatterplot of the predictor, Tanner Score, and the criterion, Frequency of Aggressive Acts.

The regression equation estimated for the data in Figure 6.4 is

$$PAAP87 = 13.43 + 0.30 * T + Residual.$$

The statistical relationship between $PAAP87$ and T is not significant ($t = 0.94; p = 0.35$).

The left panel in Figure 6.4 illustrates that the variation around the regression line, that is, the error variance, increases with the level of the predictor. For small Tanner scores we notice a relatively small error variance, that is, the residual scatter is very close to the regression line. As Tanner scores increase, the error variance increases as well, in other words, scatter of the residuals from the regression line increases. The residual plot for this regression analysis appears in the right- hand panel of Figure 6.4. Section 5.2 described measures for remedying problems of this type.

6.2   Residuals and Variable Relationships

The two panels of Figure 6.1 illustrate that any type of regression function can depict only a certain type of variable relationship. In Figure 6.1 the portion of variance that can be captured using straight regression lines is minimal (3.6%), although it is statistically significant. The residual plot (predictor x residuals) looks almost like the original data plot. It suggests that there is a substantial portion of systematic variation that can be explained using curvilinear regression.

We make two attempts to capture the curvilinear portion of the variation of the Dart Throw data. The first is to fit a quadratic polynomial. Results of this attempt appear in Figure 6.5.

The curve in Figure 6.5 suggests a relatively good fit. However, there is still a systematic portion of the data unexplained. The figure suggests that, through Stress Level 4, data points are about evenly distributed to both sides of the quadratic curve. Between Stress Levels 4 and 6, the majority of the data points are located below the regression line. Beyond Stress Level 6, most of the data points are located above the regression line. Only for the highest stress levels are data points below the regression line again.

Now, one way to further improve the model is to include a third- or even a higher degree polynomial, as will be explained in Section 7.2. Alternatively, one can look for a transformation of the predictor variable that gives a good model fit. For example, the scatter of the points for this example bears some similarity to a segment of the sine (or possibly a cosine) curve. Therefore, a sine or a cosine transformation of the predictor

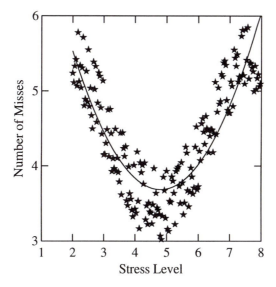

Figure 6.5: Quadratic regression of number of misses onto stress level.

may be a sensible alternative to fitting a higher degree polynomial.

To capture this portion of the data also, we substituted a sine function for the quadratic curve. Specifically, we fit the function Number of Misses = Constant + sin(Stress Level). The following estimates resulted:

Number of Misses = 4.44 + 0.99 ∗ sin (Stress Level) + Residual.

The fit for this function is as perfect as can be. The significance tests show the following results: $F_{1,178} = 961.83, p < 0.01$. The residual plot for this function appears in Figure 6.6.

Figure 6.6 suggests that, across the entire range of sin (Stress Level), residuals are about evenly distributed to both sides of the regression line. There is no systematic portion of the variation of the Number of Misses left to be detected (cf. Figure 6.2, right panel).

One word of caution before concluding this section. What we have performed here is often criticized as data fitting, that is, fitting curves to data regardless of theory or earlier results. Indeed, this is what we did. However, it was our goal to illustrate (1) how to explain data and

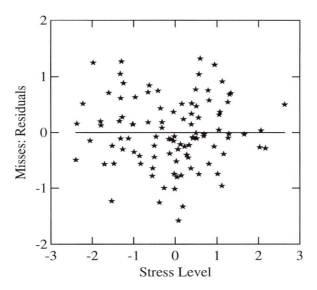

Figure 6.6: Residual misses after sine transformation of the predictor, Stress Level.

(2) what criteria to use when discussing residual plots. In real life data analysis researchers are well advised to use theory as a guide for selecting functions to fit to data.

The following paragraphs present three methods for analysis of the distribution of residuals:

1. Statistical analysis of the distribution of standardized residuals.

2. Comparison of calculated vs. expected residuals.

3. Normal probability plots.

Analysis of the Distribution of Standardized Residuals

In addition to asking whether a particular type of function best fits the data, one can analyze the distribution of residuals. If the systematic portion of the criterion variation was covered by a regression model, there is only random variation left. As a result, residuals are expected to be normally distributed. Any deviation from a normal distribution may suggest either the existence of variance that needs to be explained or the presence of outliers or both. In the following sections we describe methods for analyzing the distribution of outliers.

Consider the following regression equation:

$$y_i = b_0 + \sum_{j>0} b_j x_{ij} + e_i.$$

For the analysis of the distribution of residuals we ask whether the appropriate portion of residuals, e_i, is located at normally distributed distances from the Expectancy of e_i, $E(e_i) = 0$. For instance, we ask whether

- 68% of the residuals fall between $z = -1$ and $z = +1$ of the standard normal distribution of the residuals;

- 90% fall between $z = -1.64$ and $z = +1.64$;

- 95% fall between $z = -1.96$ and $z = +1.96$;

- 99% fall between $z = -2.58$ and $z = +2.58$; and so forth.

To be able to answer this question, we need to standardize the residuals. The standardized residual, z_{e_i}, can be expressed as

$$z_{e_i} = \frac{e_i}{s_e}, \tag{6.1}$$

where

$$s_e = \sqrt{\frac{\sum_i e_i^2}{n - p}}. \tag{6.2}$$

Using these terms, determining whether residuals are normally distributed can be performed via the following three steps:

- Calculate for each case i the standardized residual, z_{e_i};

- Count the number of cases that lie within *a priori* specified boundaries;

- Determine whether the number of cases within the a priori specified boundaries deviates from expectation; this step can be performed using χ^2 analysis.

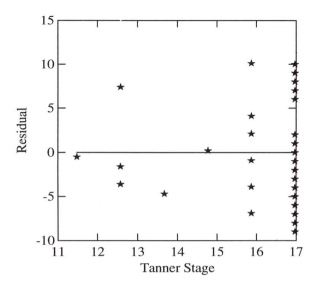

Figure 6.7: Residual plot for regression of Aggressive Acts on Tanner score.

For the following example consider the $n = 40$ boys of the Finkelstein et al. (1994) study that were still available in 1987 for assessment of pubertal status, Tanner Stage (*T87*), and Frequency of Aggressive Acts against Peers (*A87*). Regressing *A87* onto *T87* yields the following parameter estimates:

$$A87 = 1.59 + 1.10 * T87 + \text{Residual}.$$

The relationship between these two variables is statistically not significant ($t = 1.746; p = 0.09$). The residual plot appears in Figure 6.7.

The figure suggests that the size of residuals increases with the level of the predictor, Tanner Stage (cf. Figure 6.4). For the present purposes, however, we focus on the distribution of the residuals.

Table 6.1: *Raw Scores and Residuals for Aggression Study*

T87	PAAP87	Estimate	Res	Res^2	Std. Res
14	8	16.97	−8.97	80.49	−1.68
14	8	16.97	−8.97	80.49	−1.68
14	9	16.97	−7.97	63.54	−1.49
14	10	16.97	−6.97	48.60	−1.31
13	9	15.87	−6.87	47.24	−1.29
14	11	16.97	−5.97	35.66	−1.12
14	12	16.97	−4.97	24.72	−0.93
11	9	13.68	−4.68	21.86	−0.88
14	13	16.97	−3.97	15.77	−0.74
13	12	15.87	−3.87	15.00	−0.73
10	9	12.58	−3.58	12.79	−0.67
14	14	16.97	−2.97	8.83	−0.56
14	14	16.97	−2.97	8.83	−0.56
14	14	16.97	−2.97	8.83	−0.56
14	15	16.97	−1.97	3.89	−0.37
14	15	16.97	−1.97	3.89	−0.37
10	11	12.58	−1.58	2.49	−0.30
14	16	16.97	−0.97	0.94	−0.18
14	16	16.97	−0.97	0.94	−0.18
13	15	15.87	−0.87	0.76	−0.16
9	11	11.48	−0.48	0.23	−0.09
14	17	16.97	0.03	0.00	0.01
12	15	14.77	0.23	0.05	0.04
14	18	16.97	1.03	1.06	0.19
14	18	16.97	1.03	1.06	0.19
14	18	16.97	1.03	1.06	0.19
14	18	16.97	1.03	1.06	0.19
14	19	16.97	2.03	4.11	0.38
13	18	15.87	2.13	4.52	0.40
13	18	15.87	2.13	4.52	0.40
13	20	15.87	4.13	17.03	0.77
14	23	16.97	6.03	36.34	1.13
14	23	16.97	6.03	36.34	1.13
14	23	16.97	6.03	36.34	1.13
14	24	16.97	7.03	49.40	1.32
10	20	12.58	7.42	55.10	1.39

continued on next page

T87	PAAP87	Estimate	Res	Res^2	Std. Res
14	25	16.97	8.03	64.46	1.50
14	26	16.97	9.03	81.51	1.69
14	27	16.97	10.03	100.57	1.88
13	26	15.87	10.13	102.56	1.90

Table 6.1 contains the Tanner scores, the aggression frequencies, the estimates from the above regression equation, the residuals and their squares, and the standardized residuals, calculated using (6.1) and (6.2).

The $n = 40$ cases in Table 6.1 are rank ordered according to the size of their standardized residual, beginning with the largest negative residual. The counts are as follows:

- There are 25 z values less than ± 1;

- There are 15 z values between $|1|$ and $|1.96|$.

To determine whether this distribution fits what one would expect for 40 normally distributed values, we apply the Pearson X^2 test. The arrays of observed and expected frequencies appear in Table 6.2.

The X^2 for the frequencies in Table 6.2 is $X^2 = 4.61$ ($df = 2; p = 0.10$). This value suggests that the observed distribution of standardized residuals does not deviate significantly from the expected distribution. We therefore conclude that the linear regression model was appropriately applied to these data.

Another motivation for standardization of raw residuals comes from the fact that raw residuals do not have standard variance. Lack of constant variance is, in itself, not a problem. However, comparisons are easier when scores are expressed in the same units.

Table 6.2: *Observed and Expected Frequencies of Residual z Values*

	Frequencies		
	$|z| \leq 1$	$1 < |z| \leq 1.96$	$1.96 < |z|$
Observed	25	15	0
Expected	27.2	10.4	2.4

As an alternative to standardizing, researchers often studentize raw residuals, that is, standardize with respect to the t distribution (for examples see, for instance, Neter et al., 1996). Studentization is often preferred when sample sizes are relatively small. However, conclusions drawn from inspection of standardized residuals and studentized residuals rarely differ. Therefore, studentization will not be presented here in more detail.

Comparison of Calculated and Expected Residuals

A more detailed analysis of the distribution of residuals can be performed by comparing the calculated and the expected residuals. The expected residuals are estimated under the assumption that residuals are normally distributed. The two arrays of calculated and expected residuals can be compared by correlating them, by employing an F test of lack of fit (Neter et al., 1996) and by visual inspection (see below).

Expected residuals can be estimated as follows:

$$E(e_i) = \sqrt{\frac{\sum_i e_i^2}{n - p}} \; z \left(\frac{i - 0.375}{n + 0.25} \right). \tag{6.3}$$

To estimate the expected residuals, the calculated residuals must be rank ordered in ascending order. Beginning with the smallest residual, $i = 1$, the $E(e_i)$ can be determined. For the sake of efficiency notice that the number of expected residuals that needs to be calculated will never exceed $n/2$. The reason is that, for even sample sizes n, only the first half of the $E(e_i)$ need to be calculated. The second half mirrors the first around 0. For odd sample sizes n one can set the middle $E(e_i)$, where $i = n/2 + 0.5$, to $E(e_i) = 0$ and proceed with the remaining residuals as with even sample sizes.

To evaluate the degree to which the expected residuals parallel the calculated ones, one can apply the standard Pearson correlation. If the correlation is $r \geq 0.9$, one can assume that deviations from normality are not too damaging.

Table 6.3: *Comparing Calculated and Expected Residuals for Data from Aggression Study*

T87	PAAP87	Res.	Exp. Res.	i	z(...)	z
14	8	-8.97	-10.94	1.00	0.02	-2.05
14	8	-8.97	-9.34	2.00	0.04	-1.75
14	9	-7.97	-7.90	3.00	0.07	-1.48
14	10	-6.97	-7.15	4.00	0.09	-1.34
13	9	-6.87	-6.57	5.00	0.11	-1.23
14	11	-5.97	-5.77	6.00	0.14	-1.08
14	12	-4.97	-5.28	7.00	0.16	-0.99
11	9	-4.68	-4.70	8.00	0.19	-0.88
14	13	-3.97	-4.32	9.00	0.21	-0.81
13	12	-3.87	-3.79	10.00	0.24	-0.71
10	9	-3.58	-3.42	11.00	0.26	-0.64
14	14	-2.97	-2.94	12.00	0.29	-0.55
14	14	-2.97	-2.67	13.00	0.31	-0.50
14	14	-2.97	-2.19	14.00	0.34	-0.41
14	15	-1.97	-1.92	15.00	0.36	-0.36
14	15	-1.97	-1.49	16.00	0.39	-0.28
10	11	-1.58	-1.23	17.00	0.41	-0.23
14	16	-0.97	-0.80	18.00	0.44	-0.15
14	16	-0.97	-0.53	19.00	0.46	-0.10
13	15	-0.87	-0.16	20.00	0.49	-0.03
9	11	-0.48	0.16	21.00	0.51	0.03
14	17	0.03	0.53	22.00	0.54	0.10
12	15	0.23	0.80	23.00	0.56	0.15
14	18	1.03	1.23	24.00	0.59	0.23
14	18	1.03	1.49	25.00	0.61	0.28
14	18	1.03	1.92	26.00	0.64	0.36
14	18	1.03	2.19	27.00	0.66	0.41
14	19	2.03	2.67	28.00	0.69	0.50
13	18	2.13	2.94	29.00	0.71	0.55
13	18	2.13	3.42	30.00	0.74	0.64
13	20	4.13	3.79	31.00	0.76	0.71
14	23	6.03	4.32	32.00	0.79	0.81
14	23	6.03	4.70	33.00	0.81	0.88
14	23	6.03	5.28	34.00	0.84	0.99
14	24	7.03	5.77	35.00	0.86	1.08

continued on next page

T87	PAAP87	Res.	Exp. Res.	i	z(...)	z
10	20	7.42	6.57	36.00	0.89	1.23
14	25	8.03	7.15	37.00	0.91	1.34
14	26	9.03	7.90	38.00	0.93	1.48
14	27	10.03	9.34	39.00	0.96	1.75
13	26	10.13	10.94	40.00	0.98	2.05

Table 6.3 displays results of the calculations for the expected residuals. The first three columns in this table are taken from Table 6.1. The last four columns were created to illustrate calculations. The rows of Table 6.3 are sorted in ascending order with residual size as the only key. Counter i in Formula (6.3) appears in the second of the four right-hand columns. The third of these columns lists the values of the expression in parentheses in Formula (6.3). These values are areas under the normal curve. The last of these columns lists the z values that correspond to these areas. These are the values for the terms in parentheses in Formula (6.3). The values for the expected residuals, $E(e_i)$, appear in the first of these four columns.

Correlating the calculated residuals from Table 6.3 (termed "Res." in the header of the table) with the expected residuals yields a Pearson correlation of $r = 0.989$.[3] This seems high enough to stay with the assumption of parallel calculated and expected residuals.

Normal Probability Plots

Normal probability plots provide a graphical representation of how close the expected and the calculated residuals are located to each other. Part of many graphics packages (for example, SYSTAT and SPSS for Windows), normal probability plots represent a scatterplot of expected and calculated residuals. The points in this scatterplot follow a straight diagonal line if the calculated residuals are perfectly normal. Deviations

---

[3]For the present purposes we focus on the size of correlation rather than on significance testing. Raw correlation coefficients can be squared and then indicate the portion of variance shared in common. This information seems more important here than statistical significance.

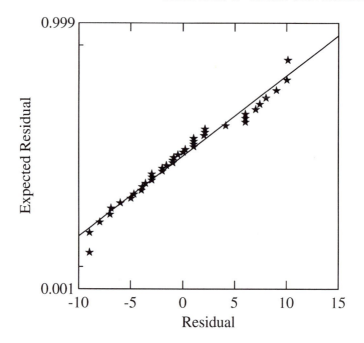

Figure 6.8: Normal probability plot for data from aggression study.

from normality result in deviations from this straight line. The plot allows researchers to exactly identify where in the distribution of residuals
deviations from normality are greatest.

For the following example we use the same data as in the last example.
Specifically, we plot the estimated residuals against the calculated ones,
both from Table 6.3. This plot appears in Figure 6.8.

Figure 6.8 displays the normal probability plot for the Tanner Stage
– Aggression residuals. The plot suggests that the coordinates of the
expected and calculated residuals very nearly follow a straight line. This
line is inserted in the figure as the regression line for the regression of the
calculated residuals on the expected residuals.

Chapter 7

POLYNOMIAL REGRESSION

7.1 Basics

In many instances, relationships between predictors and criteria are known to be nonlinear. For instance, there is a nonlinear relationship between pitch and audibility, between activation level and performance (Yerkes & Dodson, 1908), between the amount of thrill and experienced pleasure, and between speed driven and fine handed. Figure 7.1 depicts the Yerkes–Dodson Law. The law states that, for a given task difficulty, medium activation levels are most inducive for performance. The law suggests that performance increases with activation level. However, as soon as activation increases past a medium level, performance decreases again.

Using one straight regression line, researchers will not be able to validly model this law. The regression line will be horizontal, as the thin line in the figure, and not provide any information about the (strong) relationship between the predictor, Activation, and the criterion, Performance.

There are two solutions to this problem. One is called *nonlinear regression*. This approach involves fitting a nonlinear function $f(x; \beta_0, \beta_1)$ to the data, for instance, $f(x; \beta_0, \beta_1) = 1 - exp(-\beta_1(x - \beta_0))$. Because of the nonlinearity of these functions in the parameters, the procedures for parameter estimation, and hypothesis testing and the construction of

Performance

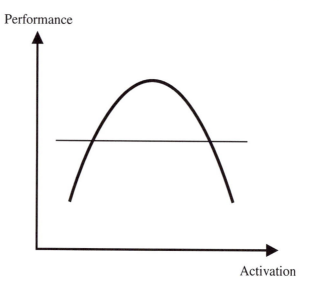

Activation

Figure 7.1: Yerkes–Dodson Law of the dependence of performance on activation.

confidence intervals do not carry over in a simple way to this situation. Parameter estimates are no longer given explicitly but must be determined iteratively, some distributional results hold only asymptotically, and so on.

However, polynomial regression is a way that allows one to fit to the data all nonlinear functions in the predictor that can be expressed as polynomials while retaining all that has been said about linear regression so far. Polynomials can be described, in general, as the product sum of parameters and x values, raised to some power,

$$
\begin{aligned}
y &= b_0 + b_1 x + b_2 x^2 + \cdots + b_J x^J \\
&= \sum_{j=0}^{J} b_j x^j,
\end{aligned}
$$

where the b_j are the polynomial parameters, and the vectors $\mathbf{x}$ contain x values, termed polynomial coefficients. The reason for using polynomial coefficients is that while we usually consider the response Y as a function

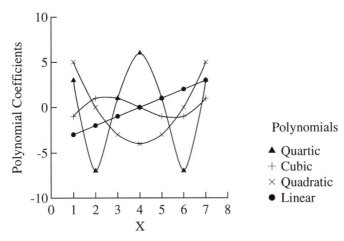

Figure 7.2: Forms of polynomials up to the fourth degree.

of the predictor X, when parameter estimation is concerned we consider the regression line as a function of the parameters. In this aspect, the previous polynomial has the same form as that of a multiple linear regression equation. While with polynomial regression not all the freedom of fitting general nonlinear models is obtained the approach is flexible enough to model a wide range of nonlinear relations without further technical complexities.

The highest power to which an x value is raised determines the degree of the polynomial, also known as the order of the polynomial. For example, if the highest power to which an x value is raised is 4, the polynomial is called a fourth-degree polynomial or fourth-order polynomial. If the highest power is J, the polynomial is a Jth-order polynomial. The form of a polynomial depends on its degree. To give an impression of possible shapes of polynomials with degrees of two, three, and four, consider Figure 7.2.

First- and second-order polynomials do not have inflection points; that is, they do not change direction. "Changing direction" means that the curve changes from a curve to the right to a curve to the left and vice versa. A look at Figure 7.2 suggests that neither the first-order polynomial nor the second-order polynomial changes direction in this sense. In contrast, third- and higher-order polynomials do have inflection points.

For example, the third-order polynomial changes direction at $X = 4$ and $Y = 0$.

The size of the parameter of the third-order polynomial indicates how tight the curves of the polynomial are. Large parameters correspond with tighter curves. Positive parameters indicate that the last "arm" of the polynomial goes upward. This is the case for all polynomials in Figure 7.2. Negative parameters indicate that the last "arm" goes downward.

The fourt-order polynomial has two inflection points. Its curve in Figure 7.2 has two inflection points, at $X = 2.8$ and $X = 5.2$, both at $Y = 0$. The magnitude of the parameter of fourth- order polynomials indicates, as for the other polynomials of second and higher order, how tight the curves are. The sign of the parameter indicates the direction of the last arm, with positive signs corresponding to an upward direction of the last arm.

Consider again the example of the Yerkes–Dodson Law. Theory dictates that the relationship between Activation and Performance is inversely U-shaped.

Figure 7.2 suggests that with only six coordinates one can create a reasonable rendering of what the Yerkes–Dodson Law predicts: At the extremes, Performance is weak. Closer to the middle of the Activation continuum, Performance increases. The simplest polynomial having such characteristics is one of second degree, that is,

$$y = b_0 + b_1 x + b_2 x^2.$$

To estimate the unknown parameters using OLS we set up the design matrix

$$\mathbf{X} = \begin{pmatrix} 1 & x_{12} & x_{12}^2 \\ 1 & x_{22} & x_{22}^2 \\ \vdots & \vdots & \vdots \\ 1 & x_{n2} & x_{n2}^2 \end{pmatrix} \qquad (7.1)$$

and perform a multiple regression analysis. From R^2 it can be seen how well the model fits, and from the scatterplot of Y against X including the fitted second degree polynomial it can be seen whether the selected polynomial gives a good description of the relation.

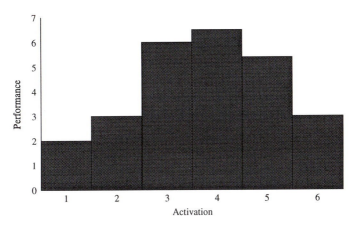

Figure 7.3: Graphical representation of the data from Table 7.1.

In the following data example we investigate whether the performance measures, obtained at six equidistant levels of Activation in a Trivial Pursuit Task, can be fit using a quadratic polynomial. Figure 7.3 displays the empirical values. Table 7.1 presents the raw data.

Taking one look at Figure 7.3 it is clear that a linear regression model is not appropriate. Nevertheless, the results of this analysis are given for completeness in Table 7.2.

The coefficient of determination is merely $R^2 = 0.131$. The slope coefficient is far from being significant. This means that a simple linear regression model can virtually explain nothing.

Table 7.1: *Raw Data for Test of Yerkes–Dodson Law Using Six Activation Levels*

Activation Level	Performance
1	2
2	3
3	6
4	6.5
5	5.4
6	3

Table 7.2: *Results of Simple Linear Regression Analysis of Activation Data*

Coefficients	Value	Std. Error	t Value	p Value
b_0	3.047	1.821	1.673	0.170
b_1	0.363	0.468	0.776	0.481

On the other hand, performing standard OLS regression analysis re-gressing Performance on a second-order polynomial in Activation level suggests a good model fit. Specifically, we calculate

$$\hat{y} = -2.52 + 4.54x - 0.60x^2,$$

where y denotes Performance, x Activation, and x^2 the squared Activation level. The results of the regression analysis are given in Table 7.3.

The coefficient of determination is $R^2 = 0.89$. The one–sided t test for the quadratic term in the model is significant at the 5% level, that is, $p = 0.0105$. We have performed the one-sided test because we hypothe-sized an inversely U-shaped relation and therefore expected the regression coefficient for the squared Activation level to be negative, $H_1 : \beta_2 < 0$. This test can be interpreted as the test of whether allowing for curvature in the regression equation improves the model fit above what would have been obtained by using only a simple linear regression model. Note that the test for the linear term, b_1, has now become statistically significant.

When selecting the polynomial to fit to a given data set, researchers need to respond to two criteria. The first is that the polynomial should, whenever possible, reflect theory. In other words, specification of shape of polynomial and number of inflection points should be based on theoretical

Table 7.3: *Quadratic Polynomial Regression Analysis of Activation Data*

Coefficients	Value	Std. Error	t Value	p Value
b_0	-2.520	1.469	1.715	0.185
b_1	4.538	0.961	4.721	0.018
b_2	-0.596	0.134	-4.437	0.021

Table 7.4: *Cubic Polynomial Regression Analysis of Activation Data*

Coefficients	Value	Std. Error	t Value	p Value
b_0	0.700	2.505	0.279	0.806
b_1	0.487	2.855	0.171	0.880
b_2	0.745	0.914	0.816	0.500
b_3	-0.128	0.086	-1.480	0.277

considerations or earlier results.[1] The second criterion concerns the order of the polynomial. Scientific parsimony requires that the polynomial be of lowest possible order.

There is, in addition, a formal constraint on the order of the polynomial to be fit to empirical data. This order cannot be greater than $t - 1$, where t is the number of different values of the predictor.

In the last example we had six different predictor values, so we could have fitted a fifth-order polynomial. As the scatterplot of Performance and Activation with the regression curve is quite acceptable, we do not expect to improve the model fit by using a third- or even fourth-degree polynomial. Nevertheless, we fit a third-degree polynomial to see whether any improvement can be achieved; that is, we use the model

$$\hat{y} = b_0 + b_1 x + b_2 x^2 + b_3 x^3.$$

We solely append a column with the Activation level in the third power to the design matrix $\mathbf{X}$ and do a multiple linear regression. It is common practice that if a kth order term is included in the model all lower order terms should be included as well and should stay in the model regardless of whether the t tests for the corresponding coefficients were significant or not. Although there may be exceptions to this rule we do not further comment on this. For a discussion of this point, see McCullagh and Nelder (1989, pp. 69). The results are given in Table 7.4.

The coefficient of determination is now $R^2 = 0.945$. It has increased

---

[1] This does not mean that polynomial approximation is not applicable in exploratory research. However, testing whether one particular polynomial provides a satisfactory rendering of a given data set presupposes theory-guided selection of polynomial.

6% over the model without the third-order term. Table 7.4 shows some very interesting facts. First of all, while the increase in R^2 suggests a third-order term may be valuable for the model, the t test of its coefficient is far from being significant. As is well known from multiple regression analysis, when adding predictors to the equation, R^2 always increases. What is perhaps the most striking feature in Table 7.4 is that the t tests for the quadratic and for the linear coefficient are now far from significant. This contradicts the earlier finding that a quadratic term will considerably improve model fit. The reason for this is that the estimated standard errors for these terms have dramatically increased. This is due to the high intercorrelations (or more generally to the almost complete linear dependency) between the X values of the first-, second- and third-order terms in the regression equation. The only reliable test is the test for the highest order term. As this test is not significant we conclude that a second-degree polynomial is adequate.

For social science data it often suffices to fit a polynomial of second or third degree as there is usually considerable variation in the data that does not allow one to determine the functional relation between Y and X more precisely. As in the example this can be done by fitting two or three different regression models of various degrees and then using the t test to find the highest order term for which the model is adequate. But sometimes the situation is not so simple, that is, higher order terms are needed to obtain an adequate model fit. In these cases the procedure suggested so far would be very laborious and time consuming and the fitting of many different regression models will often lead very quickly to confusion concerning the results and their interpretation. In these cases orthogonal polynomials are a good alternative.

7.2   Orthogonal Polynomials

Orthogonal polynomials can be considered as a technical device to overcome the problems mentioned in the last paragraph. As before, regression using orthogonal polynomials tries to fit a polynomial regression curve to the data and obtains exactly the same R^2 as using the polynomial regression approach of the last section. We will illustrate this using the example from the last section. As should be recalled, the main reason for

Table 7.5: *Polynomial Coefficients for First-, Second-, Third-, and Fourth–Order Polynomials for Six Values of X*

| Predictor | Polynomials (Order) | | | |
Values	First	Second	Third	Forth
1	−5	5	−5	1
2	−3	−1	7	−3
3	−1	−4	4	2
4	1	−4	−4	2
5	3	−1	−7	−3
6	5	5	5	1

the unreliability of the t tests other than that for the highest order term in the regression model is the high degree of linear dependency among the predictors. This phenomenon is also known as multicollinearity (see Chapter 8). In the case where the predictor is random, this is equivalent to saying that the predictors are highly intercorrelated, thus containing nearly the same information. Before going into more technical details we will analyze the foregoing example using orthogonal polynomials. If the values of the predictor are equally spaced and the number of observations belonging to each predictor value is the same, the values of the orthogonal polynomials can simply be read off of a table. In our example the predictor values are equally spaced and to each predictor value belongs a single observation. With six different predictor values the orthogonal polynomials up to the fourth degree are given in Table 7.5.

Textbooks of analysis of variance contain tables with polynomial coefficients for equally spaced predictors that typically cover polynomials up to fifth order and up to 10 different values of X (Fisher & Yates, 1963; Kirk, 1995).

Instead of using the design matrix $\mathbf{X}$ with the original predictor values in the second column (recall the column of ones in the design matrix) we replace the X values by the values of the column labeled "First" in Table 7.5. Likewise the squared predictor values in the third column of the design matrix are replaced by the column labeled "Second" in Table 7.5, and so on. This may seem quite surprising at first sight since the values in the table have obviously nothing to do with the observed Ac-

tivation levels. That this is nevertheless possible is due to the fact that
the predictor values are equally spaced. Note that the values of the column labeled "First" in Table 7.5 are equally spaced as well. Thus, these
values can be considered a linear transformation of the original Activation levels. Recall that linear transformations will possibly change the
magnitude of the corresponding regression coefficient, but its significance
remains unchanged. While the second order column of Table 7.5 is not a
linear transformation of the squared activation levels, the second-degree
polynomial regression model is equivalent to a model including a column
of ones and the first- and second order columns of Table 7.5. This carries
over to higher order polynomial models. To be more specific, the design
matrix that is equivalent to the third-degree polynomial model of the last
section is

$$
\tilde{X} = \begin{pmatrix}
1 & -5 & 5 & -5 \\
1 & -3 & -1 & 7 \\
1 & -1 & -4 & 4 \\
1 & 1 & -4 & -4 \\
1 & 3 & -1 & -7 \\
1 & 5 & 5 & 5
\end{pmatrix}.
$$

The tilde above the X indicates that it is different from the design
matrix X used earlier. Now, we perform a multiple linear regression
using this design matrix and obtain the results given in Table 7.6.

First we note that the coefficient of determination is as before $R^2 =$
0.945. In addition, the t value for the third-order term and therefore its
p value have not changed. This suggests that we are actually doing the

Table 7.6: *Cubic Regression Analysis of Activation Data Using Orthogonal
Polynomials*

Coefficients	Value	Std. Error	t Value	p Value
b_0	4.317	0.284	15.217	0.004
b_1	0.181	0.083	2.185	0.161
b_2	−0.398	0.076	−5.244	0.035
b_3	−0.077	0.052	−1.480	0.277

same test as before with the conclusion that this term adds nothing to the model. What has changed is the significance test of the second order term. Now, this test suggests, as expected, that this term contributes significantly to the model fit. In addition, the linear term is also not significant. Therefore, we have obtained the same conclusions as before but we have fitted only a single regression model rather than three different models, that is, a simple linear regression and a second- and third-order polynomial regression model. Of course, what would be worthwhile after the appropriate model has been selected is to fit this model to obtain, for example, the corresponding R^2. This can be done by dropping the third column from $\tilde{\boldsymbol{X}}$.

From this analysis it can be seen that the use of orthogonal polynomials in regression analysis can considerably simplify the analysis by drawing all important conclusions from a single analysis. This is particularly useful when the degree of the polynomial needed gets higher.

Before doing another analysis using orthogonal polynomials in which all the restrictions on the predictor values are abandoned, for example, the predictor values need no longer be equally spaced, we give an explanation as to why this simplification is achieved when using orthogonal polynomials.

Recall from Appendix A that two vectors $\mathbf{x_1}$ and $\mathbf{x_2}$ are said to be orthogonal if their inner product $\mathbf{x_1'x_2}$ is zero. As the name already suggests this is the case for orthogonal polynomials. Indeed every polynomial column in Table 7.5 is orthogonal to any other polynomial column in this table. This includes the column of ones usually added as the first column in the design matrix. This can be checked by simply calculating the inner product of any two polynomial columns in Table 7.5. These columns enter into the design matrix used in multiple linear regression. Therefore the matrix $\boldsymbol{X'X}$ is considerably simplified as it contains as its elements the inner products of all pairs of column vectors of the design matrix. In this case all off-diagonal elements of $\boldsymbol{X'X}$ are zero and therefore the inverse of $\boldsymbol{X'X}$ is readily found when calculating the OLS estimators of the parameters. Recall that

$$b = (\boldsymbol{X'X})^{-1}\boldsymbol{X'y}.$$

That OLS estimates can be easily calculated is important from a computational viewpoint. But there is another important application of the inverse of $X'X$. As outlined in Chapter 3 on multiple regression, the parameter estimates have a multivariate normal distribution with a covariance matrix given by

$$V(\mathbf{b}) = \sigma^2 (X'X)^{-1}.$$

As this matrix is diagonal, all covariances between two different elements of the **b** vector are zero; hence they are statistically independent. Therefore, when we add further columns with higher order terms to the design matrix, the parameter estimates obtained so far will not change as long as the added columns are orthogonal to all other columns already in the design matrix. This can be checked by dropping the fourth column of $\tilde{X}$ and fitting a multiple regression model with this new design matrix. The coefficients for the intercept, and the linear and second-order term will not change.

The general strategy when using orthogonal polynomials in regression analysis is therefore to fit a polynomial with a degree that is expected to describe the data a little bit too well and then look at the t values of the regression coefficients to see which degree will be sufficient to describe the data. This is now outlined in the following section.

7.3 Example of Non-Equidistant Predictors

Again consider the Yerkes–Dodson Law. While in the last example we had only six different observations and the predictor variable was assumed to be under the control of the experimenter, we now have data available from 100 persons without controlling the predictor X (Activation); that is, we observe 100 random values of the predictor. Of course these values are not equally spaced and we perhaps observe certain predictor values more often then others. The data are plotted in Figure 7.4.

The plot is again inversely U-shaped, but between the Activation levels of 2 and 6 the plot suggests that Performance is nearly constant. If we fit a second-degree polynomial to the data, using orthogonal polynomials or the approach of the last section, and include the fitted values in the scatterplot, we see that a second-degree polynomial is obviously not a good choice to model the data. This can be seen from Figure 7.5.

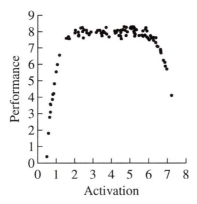

Figure 7.4: Sample of 100 Activation and Performance pairs of scores, with both the Activation and Performance variables being random.

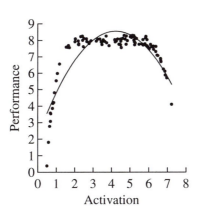

Figure 7.5: Quadratic polynomial for data in Figure 7.4.

Table 7.7: *p Values for Polynomial Terms for Data in Figure 7.4*

Coefficients	p Value
b_0	0.000
b_1	0.000
b_2	0.000
b_3	0.000
b_4	0.000
b_5	0.000
b_6	0.000
b_7	0.355
b_8	0.269

For these data to be adequately described a higher degree polynomial would be necessary. Assuming we have a computational device for obtaining orthogonal polynomials with these data, we decide to fit a model including all terms up to the eighth degree. The result of the analysis is given in Table 7.7. Because the actual values of the regression coefficient are not of interest only the p values are given.

From Table 7.7 it can be seen that a sixth-order polynomial should be selected. Figure 7.6 shows the fitted values within the scatterplot of Performance and Activation level.

As the plot suggests, the model fit is excellent. The coefficient of determination using a sixth-order polynomial is $R^2 = 0.986$. If we had not used orthogonal polynomials it would have been considerable work to find which polynomial would yielded an adequate fit.

What was left open thus far was the problem of obtaining the coefficients for the orthogonal polynomials in the general case. While for a moderate number of predictor values this could be done by hand (see, for example, Kirk, 1995, p. 761), this is not practical in the current example. For 100 observation points this would have required us to calculate polynomial coefficients to the eighth degree, that is, about 800 different values. Of course this work should be done using a computer. In S-Plus software such a function is already implemented. If the statistic software includes a matrix language, like SAS-IML, one can obtain the polynomial coefficients by implementing the Gram–Schmidt orthogonalization algo-

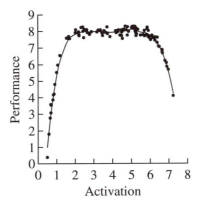

Figure 7.6: Sixth-order polynomial for data in Figure 7.4.

rithm. A description of the algorithm can usually be found in elementary textbooks of linear algebra. One starts with a design matrix, $\mathbf{X}$, for instance, Formula (7.1). That is, the design matrix contains a column of ones as the first column, the original values in the second column, the squared values in the third column, and so on. The Gram–Schmidt algorithm takes the second column vector, that is, the values of the predictor, and orthogonalizes it to the first. This is done by simply centering the predictor values. Let $\mathbf{1}$ denote a column vector of ones. The inner product of $\mathbf{1}$ and the centered predictor is,

$$\mathbf{1}'(\mathbf{x} - \bar{x}\mathbf{1}) = \sum_{i=1}^{n}(x_i - \bar{x}) = 0.$$

After the second column vector is made orthogonal to the first, the third column of the design matrix is orthogonalized to both columns preceding it, that is, the first and the just obtained second column vector. This is a bit more complex than just centering the third column, but the procedure stays straightforward.

Chapter 8

MULTICOLLINEARITY

One of the most important components of interpretation of regression parameters in multiple regression concerns the relative nature of weights. Each weight in a multiple regression equation can be interpreted as the number of steps on Y that follow one step on X, given all the other predictors in the equation. In other words, the presence of other predictors can change the parameter estimate for any given predictor. To illustrate this, we summarize parameter estimates from the three regression runs for the example from Section 3.4. In the example we predicted Breadth (Dimensionality) of Cognitive Complexity, $CC1$, from Depth of Cognitive Complexity, $CC2$, Overlap of Concepts (Categories), OVC, and Level of Education, $EDUC$. Table 8.1 presents the variable intercorrelations.

All correlations in Table 8.1 are statistically significant (even after Bonferroni adjustment for experiment-wise error; $n = 327$). Correlations in Table 8.1 suggest that Cognitive Complexity variables show high inter-

Table 8.1: *Intercorrelations of CC1, CC2, OVC, and EDUC*

	CC1	CC2	EDUC
CC2	0.154		
EDUC	0.301	0.233	
OVC	-0.784	-0.527	-0.340

correlations. In addition, Level of Education is substantially correlated with each of the Cognitive Complexity variables. The sign of the correlations are as one would expect:

1. The positive correlation between *EDUC* and *CC1* suggests that individuals with more formal education display more Breadth of Cognitive Complexity.

2. The positive correlation between *EDUC* and *CC2* suggests that individuals with more formal education have more Depth in their concepts.

3. The negative correlation between *EDUC* and *OVC* suggests that individuals with more formal education have crisper, that is, less overlapping, concepts.

As one can imagine, correlations between predictors can have effects on the size of regression parameter estimates. Thus, the presence or elimination of variables can have substantial effects on the appraisal of any β. However, this is not necessarily the case. Table 8.2 contains examples of both. The variable whose b weight is most affected by the presence of the other predictors is *EDUC*. In the unconstrained model the β estimate is 0.39. In the second constrained model, which contains only predictor *EDUC*, the estimate is 1.94.

Suppose the significance tests in the second to last column of Table 8.2 had been performed in their two-tailed form. Then, whereas the unconstrained model would have suggested that *EDUC* does not significantly contribute to predicting *CC1* ($p = 0.057$), the second constrained model would have suggested the opposite conclusion ($p < 0.01$).

Largely unaffected in spite of their high correlations with each other and other predictors (see Table 8.1), are variables *CC2* and *OVC*. Table 8.2 suggests that parameter estimates for *CC2* and *OVC*, their standard errors, and the t values for these variables remain, within very tight limits, the same regardless of whether predictor *EDUC* is in the equation or not.

In general, one calls intercorrelated predictors multicollinear. When these correlations affect estimation of regression parameters, a problem of multicollinearity exists. If this problem exists, answering the following questions can become problematic (see Neter et al., 1996, p. 285):

Table 8.2: *Three Regression Runs for Prediction of Breadth of Cognitive Complexity*

Variable	Coefficient	Std. Error	t Value	p Value (1 Tailed)
Unconstrained Model				
Intercept	29.39	1.52	19.28	< 0.01
CC2	−0.21	0.02	−10.25	< 0.01
OVC	−22.40	0.86	−26.09	< 0.01
EDUC	0.39	0.21	1.91	0.03
Constrained Model 1				
Intercept	31.40	1.10	28.45	< 0.01
CC2	−0.21	0.02	−10.10	< 0.01
OVC	−22.83	0.83	−27.44	< 0.01
Constrained Model 2				
Intercept	1.96	1.67	1.18	0.12
EDUC	1.94	0.34	5.69	< 0.01

1. What is the relative effect that each predictor has?

2. What is the magnitude of each predictor's unique effect?

3. Can predictors be eliminated because their contribution is too little?

4. Should one include additional predictors in the equation?

Answering these questions is straightforward only if the predictors in the equation are uncorrelated and, in addition, do not correlate with other variables that could possibly be used as predictors. In this case, parameter estimates remain the same regardless of what other (uncorrelated) predictor is included in the equation. If, however, and this is typical of social science empirical data, predictors are correlated, problems may occur. While multicollinearity does not, in general, prevent us from estimating models that provide good fit (see Table 8.2), interpretation of a parameter estimate becomes largely dependent upon what other predictors are part of the regression equation.

8.1 Diagnosing Multicollinearity

Among the most important indicators of multicollinearity are the following:

1. Correlations exist among predictors (see Table 8.1).

2. Large changes occur in parameter estimates when a variable is added or removed (see Table 8.2).

3. Predictors known to be important do not carry statistically significant prediction weights (see two-tailed test of *EDUC* in unconstrained model in Table 8.2).

4. The sign of a predictor is counterintuitive or even illogical.

5. Surprisingly wide confidence intervals exist for parameter estimates for predictors known to be of importance.

6. There exists a large variance inflation factor (VIF).

While many of these indicators are based on prior knowledge and theory, others can be directly investigated. Specifically, Indicators 3 and 4 are fueled by substantive knowledge and insights. All the others can be directly investigated. In the following we give a brief explanation of Indicator 6, the variance inflation factor.

Consider an unconstrained multiple regression model with p predictors ($p > 1$). Regressing predictor m onto the remaining predictors can be performed using the following regression equation:

$$X_m = b_0 + \sum_{j \in M} b_j X_j + \text{Residual}, \quad \text{for } j \neq m.$$

Let R_m^2 be the coefficient of multiple determination for this model. Then, the VIF for predictor m is defined as

$$\text{VIF}_m = \frac{1}{1 - R_m^2}. \tag{8.1}$$

The VIF has a range of $1 \leq \text{VIF} \leq +\infty$. The VIF increases with the severity of multicollinearity. Only when $R_m^2 = 0$, that is, when predictors

Table 8.3: *Regressing Predictors CC2, OVC, and EDUC onto Each Other*

Variable	Coefficient	Std. Error	t Value	p Value (1 Tailed)
Dependent Variable CC2				
Intercept	46.25	3.25	14.21	< 0.01
EDUC	0.68	0.56	1.21	0.11
OVC	-20.59	2.04	-10.11	< 0.01
Dependent Variable OVC				
Intercept	1.24	0.068	18.86	< 0.01
EDUC	-0.06	0.01	-4.90	0.01
CC2	-0.01	0.001	-10.11	< 0.01
Dependent Variable EDUC				
Intercept	5.14	0.30	17.30	< 0.01
CC2	0.01	0.01	1.21	0.11
OVC	-1.10	0.22	-4.90	< 0.01

are uncorrelated, does one obtain the minimum value, VIF $= 1$. It is important to note that the VIF is variable-specific. As was indicated in the example at the beginning of this chapter, not all variables are equally affected by multicollinearity.

The VIF increases as the multiple correlation among predictors, predicting other predictors, increases. A rule of thumb is that when the VIF ≥ 10, problems with multicollinearity are severe, that is, multicollinearity greatly influences the magnitude of parameter estimates.

To illustrate the VIF we use the example again where we predict Breadth of Cognitive Complexity (*CC1*), from Depth of Cognitive Complexity (*CC2*), Overlap of Concepts (*OVC*), and Level of Education (*EDUC*). In the following we estimate the VIF for each of the three predictors, *CC2*, *OVC*, and *EDUC*.

To do this, we have to estimate the following three regression models:

$$\text{CC2} = b_0 + b_1 * \text{OVC} + b_2 * \text{EDUC} + \text{Residual},$$
$$\text{OVC} = b_0 + b_1 * \text{CC2} + b_2 * \text{EDUC} + \text{Residual},$$
$$\text{EDUC} = b_0 + b_1 * \text{CC2} + b_2 * \text{OVP} + \text{Residual}.$$

Results of these analyses appear in Table 8.3.

The multiple correlations for the three regressions are 0.281, 0.328, and 0.119, respectively. Inserting the R^2 into Equation (8.1) yields

$$\text{VIF}_{CC2} = \frac{1}{1 - 0.281} = 1.391,$$

$$\text{VIF}_{OVC} = \frac{1}{1 - 0.328} = 1.488,$$

$$\text{VIF}_{EDUC} = \frac{1}{1 - 0.119} = 1.135.$$

These values can be interpreted as the factor by which the expected sum of squared residuals in the OLS standardized regression coefficients inflates due to multicollinearity. In other words, because the predictors are correlated, the expected sums of squared residuals are inflated by a factor of the VIF.

For example, because the three predictors $CC2$, OVC, and $EDUC$ are correlated, the expected residual sum of squares for $CC2$ is inflated by a factor of 1.391. In the present example, none of the VIF values come even close to the critical value of 10. Therefore, we can conclude that analysis of the present data does not face major multicollinearity problems. Nevertheless, notice how the sign of the regression coefficient for $CC2$ changes in the presence of $EDUC$, and how the sign of the coefficient for $EDUC$ changes depending on whether $CC2$ or OVC is in the equation (Table 8.3).

8.2   Countermeasures to Multicollinearity

There is a number of countermeasures one can take in the presence of multicollinearity. In the following paragraphs we give a brief overview of countermeasures. Examples and more detailed explanations of the effects of centering follow.

Centering Predictors

Centering predictors often reduces multicollinearity substantially. For example, squaring coefficients of a linear trend to assess a quadratic trend generates vectors that are strongly correlated with each other. Using orthogonal polynomial parameters reduces this correlation to zero. Orthogonal polynomial parameters are centered (see Section 7.2; an example of the effect created by centering predictors is given below).

Dropping Predictors

When parameter estimates for certain predictors severely suffer from multicollinearity one may drop one or more of these predictors from the multiple regression equation. As a result, inflation of standard errors of parameter estimates will be reduced. There are, however, several problems associated with this strategy. Most importantly, dropping predictors prevents one from estimating the contribution that these predictors could have made to predicting the criterion. In addition, estimates for the predictors remaining in the model may be affected by the absence of the eliminated predictors.

Ridge Regression

This method creates biased estimates for β. However, these estimates tend to be more stable than the unbiased OLS estimates (for details see Neter et al., 1996, cf. Section 12.3).

Performing Regression with Principle Components

This approach transforms predictors into principal component vectors which then are used as predictors for multiple regression. For details see Rousseeuw and Leroy (1987); for a critical discussion of this method see Hadi and Ling (1998).

8.2.1 Data Example

In the following paragraphs we illustrate the effect created by centering predictors. For an example, consider the case in which a researcher asks whether there exists a quadratic trend in a data set of five measures. To test this hypothesis, the researcher creates a coding variable that tests the linear trend, and a second coding variable that is supposed to test the quadratic trend. Table 8.4 displays the coding variables created by the researcher.

Table 8.4: *Centering a Squared Variable*

Lin Trend I	Lin Trend II	Quad Trend I	Quad Trend II
A	B	F	G
1	−2	1	4
2	−1	4	1
3	0	9	0
4	1	16	1
5	2	25	4

Table 8.4 displays four coding variables. Variable A was created to assess the linear trend. It shows a monotonic, linear increase. Variable B results from centering A, that is, from subtracting the mean of A, $\bar{A} = 3$, from each value of A. Variable F was created to assess the expected quadratic trend. Variable F results from squaring the values of A. Variable G was also created to assess the quadratic trend. It results from squaring B, the centered version of A.

As a means to determine whether using variables from Table 8.4 could cause multicollinearity problems we correlate all variables that could possibly serve as predictors in a multiple regression, that is, all variables in the table. The correlation matrix appears in Table 8.5.

Table 8.5 shows a very interesting correlation pattern. As expected, Variable A correlates perfectly with Variable B. This illustrates that the Pearson correlation coefficient is invariant against linear transformations. Both Variables A and B correlate very highly with Variable F, that is, the square of A. This high correlation illustrates that squaring variables typically results in very high correlations between the original and the

Table 8.5: *Correlation Matrix of Centered and Uncentered Predictors*

	A	B	F
B	1		
F	0.98	0.98	
G	0	0	0.19

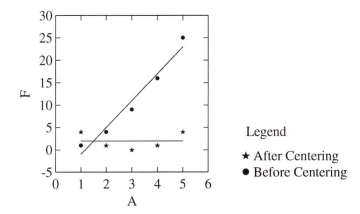

Figure 8.1: Relationships between linear and quadratic predictors before and after centering.

squared variables. However, the square of B, called G in Tables 8.4 and 8.5, correlates neither with A nor with B. It correlates only with F, the other squared variable. Figure 8.1 displays the relationships between Variables A and G and A and F.

The results from Tables 8.4 and 8.5 and Figure 8.1 can be summarized as follows: When predictors are highly correlated with other predictors because they result from multiplying predictors with themselves or each other, centering before multiplication can greatly reduce variable intercorrelations. As the present example shows, variable intercorrelations can, under certain conditions, even be reduced to zero.

Chapter 9

MULTIPLE CURVILINEAR REGRESSION

Chapter 8 introduced readers to simple curvilinear regression. This topic is taken up here again. Although researchers in the social sciences chiefly rely on describing variable relationships using straight regression lines, it is generally acknowledged that there are many applications for curvilinear relationships. Examples of such relationships include the Yerkes–Dodson Law (see Figure 7.1), learning curves, forgetting curves, position effects in learning, learning plateaus, item characteristics, and the concept of diminishing returns. In other sciences, curvilinear relationships are quite common also. For example, in physics the relationship between temperature and vapor pressure is curvilinear, and so is the function relating heat capacities and temperature to each other. Similarly, sone (loudness) scales are nonlinear. Parameters for many of these functions can be estimated using the least squares methods discussed in this volume.

The first two figures depict two scenarios of multiple regression. Figure 9.1 depicts standard *linear* multiple regression using the function $Z = X + Y$. The graph shows a straight plane, a regression hyperplane, sloped at a 45° angle. Figure 9.2 presents an example of a curvilinear regression hyperplane using the function $Z = 0.5X^2 + 0.2Y^3$. A function of

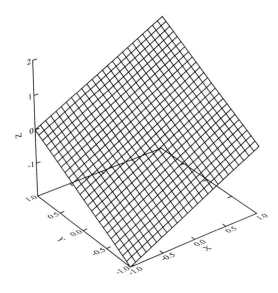

Figure 9.1: Regression hyperplane for $Z = X + Y$

this type can be estimated using ordinary least squares methods because parameters are linear, that is, raised to the first power. Predictors need not be linear.

The least squares principles for parameter optimization for curvilinear functions are the same as those for linear functions. In each case the goal is to find those coefficients that yield the smallest possible sum of squared residuals, that is, to minimize the sum of squared differences between observed values and the regression hyperplane. This is done, as was illustrated in detail in Section 2.2.2, by creating the first derivative of the function of squared residuals and setting it equal to zero. Using the terminology from Appendix C on vector differentiation we can describe this process as follows. The quadratic form to be minimized is

$$\text{SSR} = (\mathbf{y} - \mathbf{Xb})'(\mathbf{y} - \mathbf{Xb}),$$

where *SSR* is the sum of squared residuals, and the vectors and matrices

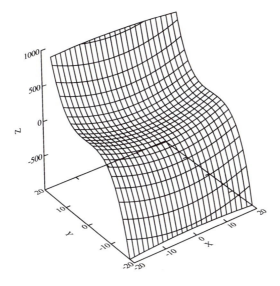

Figure 9.2: Regression hyperplane for $Z = 0.5X^2 + 0.2Y^3$

are as before. The minimization process produces

$$\frac{\partial}{\partial b_i} \text{SSR}(\mathbf{b}) = 0,$$

for all b_i where $i = 0, ..., p$, and with $p+1$ denoting the number of vectors in $\mathbf{X}$.

The following are examples of curvilinear functions:

1. Exponential Function

$$y = b_0 + b_1 exp(b_2 x)$$

2. Polynomial

$$y = \sum_{j=0}^{p} b_j x^j$$

3. Trigonometric Function

$$y = b_0 + b_1 \sin x + b_2 \cos x$$

4. Vapor Pressure Function

$$y = b_0 + \frac{b_1}{x} + b_2 \log x + b_3 x + \cdots,$$

where x is the vapor temperature (in Kelvin) and the dependent variable is $y = \log P$, the logarithm of vapor pressure

5. Heat Capacity Function

$$y = b_0 + b_1 x + \frac{b_3}{x^2}$$

6. Negatively Accelerated Function of Practice Trials (Hilgard & Bower, 1975)

$$p_n = 1 - (1 - p_1)(1 - \theta)^{n-1},$$

where n denotes the trials, θ is the flatness parameter indicating how steep the increase is in response probability (p_n), and p_1 is the probability of the first response; Figure 9.3 gives three examples of such curves.

The following example uses data from the project by Spiel (1998) on the development of performance in elementary school. We analyze the variables Fluid Intelligence F, Crystallized Intelligence C, and Grades in German G in a sample of $n = 93$ second graders. The hypotheses for the analyses are:

1. Grades in German can be predicted from a linear combination of Fluid Intelligence, that is, training-dependent Intelligence, and Crystallized Intelligence;

2. The effects of Fluid Intelligence taper off, so that the relationship between Fluid Intelligence and Grades is nonlinear; and

3. The relationship between Crystallized Intelligence and Grades is linear.

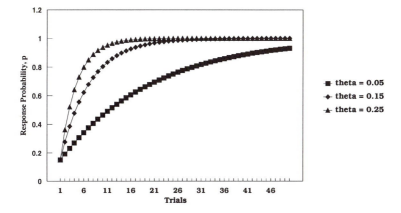

Figure 9.3: Three examples of functions of practice trials

To test this set of hypotheses we perform multiple, nonlinear regression as follows. We create a nonlinear version of variable Fluid Intelligence. For starters, we square the raw scores and obtain $F2$. The second predictor is Crystallized Intelligence. We use this variable without transforming it. From the data we estimate the following regression parameters

$$G = 4.94 + 0.11 * F2 + 0.02 * C + \text{Residual}.$$

The multiple R^2 for this equation is 0.43. The statistical analyses indicate that whereas $F2$ makes a statistically significant contribution ($t = 9.613; p < 0.01$), C does not ($t = 1.19; p = 0.239$). Figure 9.4 displays the relationship between $F2$, C, and G.

The following second example analyzes data using polynomial regression. In an experiment on the effectiveness of an antidepressive drug, it was asked whether the proportion of patients that recovered from reactive depression could be predicted from the dosage of the drug. The researchers used five doses of the drug: 1/2 a unit, 1 unit, 3/2 units, 2 units, and 5/2 units. Figure 9.5 displays the proportions of patients that recovered after taking the drug for four weeks (line with triangles, PROP) and the curves of the parameters of the first (line with circles ; P1), second (line with × signs, P2), and third (line with + signs, P3) order polynomials used to approximate the observed proportions. Proportions were

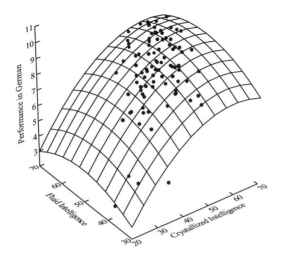

Figure 9.4: Multiple nonlinear relationships between the predictors, Crystallized Intelligence and Fluid Intelligence, and the criterion, Performance in German.

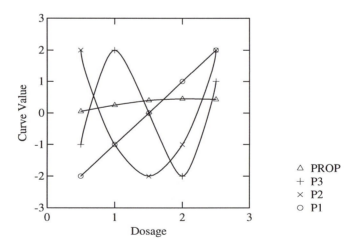

Figure 9.5: Raw scores of drug effects and indicators used to predict the raw data curve.

Table 9.1: *Raw Scores and Polynomial Coefficients for Drug Efficiency Example*

| | Observed | Polynomial Coefficients | | |
Dosage	Proportion	Linear	Quadratic	Cubic
1/2	0.05	−2	2	−1
1	0.25	−1	−1	2
3/2	0.4	0	−2	0
2	0.45	1	−1	−2
5/2	0.43	2	2	1

calculated using independent samples.

The raw scores the polynomial coefficients used to smooth the polynomials[1] appear in Table 9.1. Using these values to estimate parameters for a multiple regression equation yields

Proportion $= 0.32 + 0.096 * P1 - 0.039 * P2 + 0.002 * P3 +$ Residual.

Whereas the parameters for the linear and the quadratic component of this equation are statistically significant, the parameter for the third-order polynomial is not. More specifically, we calculate the following t values and tail probabilities: for P1: $t = 31.75$, $p = 0.02$; for P2: $t = -15.09$, $p = 0.04$; and for P3: $t = -0.661$, $p = 0.628$.

The multiple R^2 is 0.999. We thus conclude that the curve of proportions can be approximated almost perfectly from the first-, second-, and third-order orthogonal polynomials.

Because the third-order polynomial did not contribute significantly to the model–data fit, we now consider a reduced model. Specifically, we consider the model that only includes the first- and second-order polynomials. The polynomials included in the first model are orthogonal (see Section 7.2). Therefore, there is no need to recalculate the above equation to obtain parameter estimates for the reduced model. One can simply drop the term of the predictor not included in the reduced model. We

---

[1]Smoothing in Figure 9.5 was performed using the spline smoother in SYSTAT's GRAPH module.

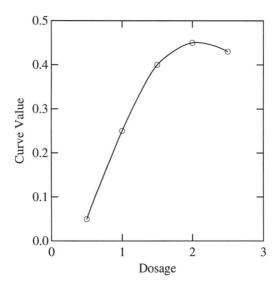

Figure 9.6: Second-order polynomial describing dosage effects ($R^2 = 0.999$).

obtain in the present example

$$\text{Proportion} = 0.32 + 0.096 * \text{P1} - 0.039 * \text{P2} + \text{Residual}.$$

The portion of variance accounted for by this model is still $R^2 = 0.999$. The test statistics for the slope parameters are: for P1: $t = 37.45$, $p = 0.001$; and for P2: $t = -17.80$, $p = 0.003$.

We thus conclude that the reduced model also provides excellent fit. To illustrate how good the fit is, consider Figure 9.6.

Figure 9.6 displays the observed proportions and the fitting curve from the first- and second-order polynomials. Obviously, the fit is close to perfect. Only for the 1 and 3/2 doses are there minor deviations.

The figure illustrates also, however, that extrapolation can suggest very misleading conclusions. Administering a dose greater than 2 units of the drug is displayed as having a lesser effect than doses just around 2 units. While this trend may be plausible within a certain range, for larger doses this would mean that proportions are negative. Readers are invited to interpret negative proportions in this particular example.

Chapter 10

INTERACTION TERMS IN REGRESSION

Regression interaction is one of the hotly and widely discussed topics of regression analysis. This chapter provides an overview of the following three topics: First, it presents a definition, along with examples, of interaction in regression. Second, it discusses the meaning of multiplicative terms in regression models. Third, it introduces the distinction between multiplicative terms and interaction terms in regression models and illustrates the use of interaction models.

10.1   Definition and Illustrations

To define interaction in regression, one needs at least three variables. Two of these are predictors and the third is the criterion in multiple regression. Consider the case where researchers predict the criterion, Y, from the predictors, X_1 and X_2, using the following regression equation:

$$Y = \beta_0 + \beta_1 X_1 + \beta_2 X_2 + \epsilon. \tag{10.1}$$

In this model, parameters β_1 and β_2 represent the regression main effects[1] of the predictors, X_1 and X_2. Interpretation of regression main effects was discussed in Section 2.3 and Section 3.2. In brief, β_1 indicates how many steps on Y follow from one step on X_1, given X_2. This applies accordingly to β_2.

Application of Equation (10.1) implies that the effects of X_1 and X_2 are independent in the sense that neither β_1 nor β_2 change with X_2 or X_1, respectively; that is, neither β_1 or β_2 are functionally related to X_2 or X_1, respectively. In other words, suppose that the first predictor is changed by one unit; then the change of the criterion is exactly β_1 regardless of the value that the second predictor currently has. An example of a regression hyperplane (also termed a response surface) for this situation is given in Figure 3.1. Noninteracting variables are also termed "additive".

There are many instances, however, where the assumption of independent regression main effects does not apply. Consider, for example, the effects of drugs taken together with alcohol. Drug effects can vary dramatically depending on blood alcohol level. In technical terms, the regression slope of drug on behavior varies with blood alcohol level. The same may apply to the effects of alcohol.[2]

Thus, regression interaction can be defined as follows: When the estimate of the slope parameter for one predictor depends on the value of another predictor, the two predictors are said to interact. There are two subtypes of regression interactions.

1. Symmetrical Regression Interaction: Two predictors mutually affect each other's slopes;

2. Asymmetrical Regression Interaction: One predictor affects the slope of the other predictor but its own slope is not affected.

When predictors interact, regression hyperplanes display curvature, indicating the changes in regression slopes. The curvature of regressions

---

[1]In the current context, we use the term "effect" in a very broad sense. When, in earlier and later chapters we say a variable allows one to predict some other variable, we associate the same meaning as when saying a variable has an effect. We use the term "effect" here for brevity of notation and because of the association to analysis of variance.

[2]In analysis-of-variance contexts, where predictor levels are categorical or categorized, this type of interaction is termed treatment-contrast interaction (see Kirk, 1995).

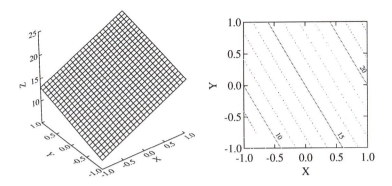

Figure 10.1: Response surface and its contours when no interaction is present.

with interactions are of a certain kind: the contour lines[3] are not parallel. In contrast, when there exists no interaction, contour lines are parallel. In order to give a visual impression of these three– dimensional surfaces we suppose for the moment that the true regression coefficients are known. Three pairs of figures illustrate this relationship. The left panel of Figure 10.1 displays the hyperplane of the multiple regression function, $Y = 15 + 5X_1 + 3X_2$. Interactions are not part of the model. Therefore, the plane is without curvature. The contour lines for this hyperplane appear in the right panel. The lines are perfectly parallel. They indicate that the hyperplane slopes upward as the values of both X_1 and X_2 increase. This happens in a perfectly regular fashion.

In contrast, Figure 10.2 displays a response surface with an interaction. Specifically, the left panel of Figure 10.2 displays the surface for the model $Y = 15 + 5X_1 + 3X_2 - 15X_1X_2$, that is, the same model as in Figure 10.1, but with an added multiplicative term. The figure indicates that Y increases with X_1 and X_2, and that the increase is negatively accelerated, that is, becomes smaller. In addition, the response surface displays smaller values as the difference between X_1 and X_2 becomes smaller.

The right panel of Figure 10.2 displays the contour lines for this re-

---

[3]Contour lines are projections of elevation on the Z-axis onto the X-Y plane. Contour lines indicate where the elevation is the same.

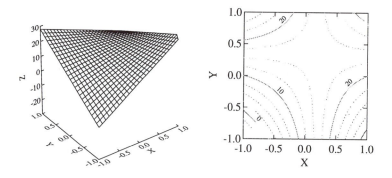

Figure 10.2: Response surface and its contours when multiplicative interaction is present.

gression model. No doubt, these lines are not parallel. This can be seen from the elevations indicated by the lines.

It is important to notice that nonlinear response surfaces do not necessarily suggest the presence of interactions. This is illustrated in Figure 10.3. The left panel of this figure displays the response surface for the model $Y = 70 - 2X_1^2 - 3X_2^2$. The contour plot for this model appears in the right panel. Obviously, the contour lines are not linear, but they are parallel. We conclude from these examples that it may not always be obvious from the $Y \times X_1 \times X_2$ plot whether there exists an interaction. The contour plot can reveal the existence of interaction. When there are more than two predictors, it may not be possible to create a visualization of variable relationships. Therefore, researchers typically focus on using statistical and conceptual approaches to discuss possible interactions. The following section introduces readers to concepts of multiplicative terms and interactions.

10.2 Multiplicative Terms

This section explains and illustrates two situations in which a regression model contains a multiplicative term. First, this chapter addresses the problem of asymmetric regression interaction, that is, the case where one predictor in multiple regression determines the slope of the second. Sec-

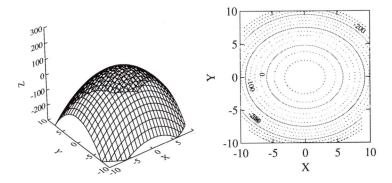

Figure 10.3: Nonlinear response surface and its contours when no interaction is present.

ond, the chapter addresses the case where one predictor determines both the intercept and the slope of a second predictor (for more in–depth coverage of these topics see Cohen, 1978; Fisher, 1988; Bryk & Raudenbush, 1992; Rovine & von Eye, 1996).

10.2.1 Predicting the Slope Parameter from a Second Predictor

Consider the case of a simple regression that relates a predictor, X_1, and a criterion, Y, to each other, or

$$Y = \beta_0 + \beta_1 X_1 + \epsilon.$$

If one adds a second predictor, X_2, to this model, one can create a standard multiple regression equation, that is,

$$Y = \beta_0 + \beta_1 X_1 + \beta_2 X_2 + \epsilon.$$

In this case, one would assume that the predictors X_1 and X_2 have independent effects upon the criterion Y.

In contrast, we now assume that X_2 has an effect on the regression slope of X_1, β_1. This can be expressed using the following system of two equations

$$Y = \beta_0 + \beta_1 X_1 + \epsilon \tag{10.2}$$
$$\beta_1 = \beta_2 + \beta_3 X_2. \tag{10.3}$$

Note that Equation (10.3) describes a functional relationship between the regression slope of the first predictor and the values of the second predictor that is linear. This should not be confused with the usual regression equation that relates a criterion to a predictor, allowing for a residual term that is supposed to be stochastic. Substituting Equation (10.3) into Equation (10.2) yields the multiple regression equation

$$Y = \beta_0 + (\beta_2 + \beta_3 X_2) X_1 + \epsilon,$$

which has the form of a simple regression of Y on X_1, but the slope now depends on X_2. Solving the middle expression yields

$$Y = \beta_0 + \beta_2 X_1 + \beta_3 X_1 X_2 + \epsilon. \tag{10.4}$$

This equation does include a multiplicative (or product) term, $X_1 X_2$. The model described by this equation assumes that the slope of the regression of Y on X_1 depends on the value of X_2. More specifically, for a discrete subset of values of X_2, this equation assumes that the slope β_1 varies monotonically with X_2. For example, it could be that the slope increases proportionally with X_2. In addition, Equation (10.4) implies that the criterion, Y, does not vary with X_2 when $X_1 = 0$.

Equation (10.4) can be estimated. If parameter estimate b_3 is statistically significant, one can assume that the effect of X_1 on Y depends on X_2, or is *mediated* by X_2. In other words, if $b_3 \neq 0$, then the effect of X_1 on Y is not constant across the values of Y_2.

Equation (10.4) illustrates that the assumption that the regression of one variable onto a second depends on a second predictor yields a regression equation with one main effect and one product term. Treating the multiple regression model given in (10.4) as the unconstrained model and the simple regression model given in (10.2) as the constrained model, one can test whether including the product term results in a significant improvement over the simple model. If the interaction accounts for a significant portion of variance, parameter estimate b_3 will be different than zero, and one can conclude that the parameter estimate, b_1, is not constant across the observed range of values of predictor X_1.

The following numerical examples use data from an experiment on recall of short narratives that differed in concreteness, *TG* (von Eye, Sörensen, & Wills, 1996). A sample of $n = 327$ adults, aged between 18 and 70, read two short narratives each. The instruction was to read and memorize the texts. Before reading the texts, each participant solved a task that allowed researchers to determine cognitive complexity, *CC1*. The dependent variable was the number of text propositions recalled, *REC*.

In the first example we illustrate use of Formulas (10.2), (10.3), and (10.4). We regress recall performance, *REC*, onto text concreteness, *TG*, and obtain the following regression equation:

$$REC = 113.39 - 25.54 * TG + \text{Residual}.$$

The negative sign for *TG* in this equation suggests that more concrete texts are better recalled than more abstract texts. The slope coefficient is statistically significant ($t = -6.27; p < 0.01$), thus suggesting that recall depends on text concreteness. The portion of criterion variance accounted for is $R^2 = 0.11$.

For the following analysis we assume that the slope parameter for text concreteness is mediated by subject cognitive complexity. Specifically, we assume that subjects differing in cognitive complexity differ in the way they exploit the recall advantage of text concreteness. From this assumption we create an equation of the form given in (10.4) and obtain the following parameter estimates:

$$REC = 112.60 - 30.24 * TG + 0.47 * (TG * CC1) + \text{Residual}.$$

Both parameters in this equation are significant ($t = -6.53, p < 0.01; t = 2.11, p = 0.04$, respectively), thus indicating that the null hypothesis of no mediation can be rejected. The portion of variance accounted for by the model with the product term is $R^2 = 0.12$.

The left panel of Figure 10.4 displays the response surface for the

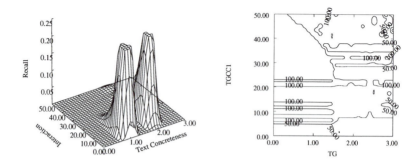

Figure 10.4: Relationship between the predictors, Text Concreteness and interaction with Cognitive Complexity, and a contour plot.

regression model that includes the interaction term.[4] The surface suggests a bimodal distribution with a higher peak for concrete texts than for abstract texts. (Intermediate texts were not used in this experiment.) This suggests that recall is better for concrete texts than for abstract texts. The figure also suggests that the variance of $(TG * CC1)$ values is greater for $TG = 2$. Thus, there may be a problem with heteroscedasticity in these data. The right panel of Figure 10.4 displays the contour plot for these data.

We now ask whether the model with the product term is statistically better than the model without. To do this, we consider the model with the product term the unconstrained model. The model without the product term is the constrained one. Inserting into Formula (3.24) of Section 3.3.1 we obtain

$$F = \frac{\frac{0.120-0.108}{2-1}}{\frac{1-0.120}{327-2-1}} = 4.418,$$

a value that is identical to the t value given above ($t = 2.11^2 = 4.45$) which is, within rounding, equal to $F = 4.418$ (this relationship between t and F always applies, when $df_1 = 1$ for the F test). For $df_1 = 1$ and

---

[4]The surface was created using SYSTAT's kernel estimation option, available in the CONTOUR routine of the GRAPH module.

Table 10.1: *Intercorrelations of Text Concreteness, TG, Recall, REC, Cognitive Complexity, CC1, and the Product Term, REC*CC1*

	CC1	TG	REC
TG	-0.071	1.000	
REC	0.119	-0.328	1.000
TG * CC1	0.792	0.483	-0.062

$df_2 = 324$ this F value has a tail probability of $p = 0.036$. Smaller than $\alpha = 0.05$, this value allows us to conclude that the more complex model, the one with the multiplicative term, explains the data better than the more parsimonious model without the multiplicative term.

Table 10.1 contains the correlations between the variables used for these analyses. Readers are invited to discuss possible multicollinearity problems present in these analyses.

10.2.2 Predicting Both Slope and Intercept

Consider the case where not only the slope but also the intercept depends on the level of a second predictor. This case can be described by the following system of three equations:

$$Y = \beta_0 + \beta_1 X_1 + \epsilon \tag{10.5}$$
$$\beta_1 = \beta_2 + \beta_3 X_2 \tag{10.6}$$
$$\beta_0 = \beta_4 + \beta_5 X_2. \tag{10.7}$$

Inserting (10.6) and (10.7) into (10.5) yields

$$Y = \beta_4 + \beta_5 X_2 + \beta_2 X_1 + \beta_3 X_1 X_2 + \epsilon. \tag{10.8}$$

Obviously, this is a multiple regression equation that includes both main effects and the multiplicative term. In this equation, the tests concerning the parameters $\beta_4, \beta_5, \beta_2$, and β_3 have a meaning different than the meaning usually associated with these tests.[5] More specifically, the

---

[5]It should be noted that one of the main differences between the model given in

following tests are involved:

- Test of β_5: is the intercept, β_0, dependent on X_2?

- Test of β_3: is the slope, β_1, dependent upon X_2?

Results of these tests can be interpreted as follows:

1. If β_5 is not zero, then the intercept of the regression of Y on X_1 depends on X_2

2. If $\beta_5 = 0$, then the relationship between β_0 and β_4 is constant, regardless of what value is assumed by X_2, and the intercept, β_0, remains the same for all values of X_2

3. If β_3 is zero, the slope of the regression of Y on X_1 is constant and unequal to zero across all values of X_2

4. If both β_2 and β_3 are zero, there is no relationship between Y and X_1

5. If β_3 is not zero, then the slope of the regression of Y on X_1 depends on X_2.

The following example uses data from the von Eye et al. (1996) experiment again. We now regress the adults' recall on their educational background *EDUC*. For the $n = 327$ participants we estimate

$$REC = 46.07 + 6.03 * EDUC + Residual,$$

which indicates that subjects with a higher education (measured as number of years of formal school training) tend to have higher recall rates. While only accounting for $R^2 = 0.020$ of the criterion variance, this relationship is statistically significant ($t = 2.60, p = 0.010$).

One may wonder whether the regression line and intercept of this relationship depend on the subjects' cognitive complexity. If this is the

---

(10.8) and the model given in (10.4) is that in (10.8) the criterion, Y, can depend on X_2 even if X_1 assumes the value 0.

Table 10.2: *Significance Tests for Prediction of Intercept and Slope Parameters of Regression of Recall on Education*

Variable	Parameter	Std. Error	t value	p value
Intercept	$b_4 = -10.78$	26.24	-0.41	0.682
CC1	$b_5 = 5.33$	2.06	2.58	0.010
EDUC	$b_2 = 16.60$	5.50	3.02	0.003
CC1 * EDUC	$b_3 = -0.97$	0.41	-2.35	0.019

case, the intercept of the slope and the slope itself may vary with subjects' cognitive complexity. We estimate the following regression parameters:

$$REC = -10.78 + 5.34 * CC1 + 16.60 * EDUC$$
$$- 0.97 * CC1 * EDUC + Residual.$$

While still only explaining 4.3% of the criterion variance, this equation contains only significant slope parameters. The intercept parameter is not significant. Table 10.2 gives an overview of significance tests.

These results can be interpreted as follows:

- The intercept of the regression of *REC* on *EDUC* is zero ($p(b_5) > 0.05$);

- However, this intercept varies with *CC1* ($p(b_4) < 0.05$)

- The slope of the regression of *REC* on *EDUC* depends on *CC1* ($p(b_2) < 0.05$ and $p(b_3) < 0.05$).

Caveats and Problems

There are two major issues that need to be considered when including multiplicative terms in multiple regression. The first of these issues concerns model specification. The second concerns characteristics of the multiplicative terms. The former issue is addressed in the following paragraphs, and the latter in Section 10.3.

Model Specification

Section 10.2 illustrated that hypotheses concerning interactions between two predictors can lead to two different types of models. The first type of model involves one main effect term and one product term. This model is asymmetric in nature in that it considers the effect of predictor P_2 on the slope of the regression line for C on P_1. Investigating the inverse effect, that is, the effect of predictor P_1 on the slope of the regression line of C on P_2, implies a different model. This model involves the same multiplicative term as the first. However, it involves the main effect of the other predictor.

In contrast, investigating a multiple regression model of the type given in Formulas (10.5), (10.6) and (10.7) implies a symmetrical model in the sense that (10.8) does not allow one to discriminate between the models that investigate

1. the effects that predictor P_2 has on the slope and intercept of the regression line for C on P_1, and

2. the effects that predictor P_1 has on the slope and intercept of the regression line for C on P_2.

Thus, researchers should be aware that always using one particular regression model to treat interaction hypotheses can lead to misspecified models and, therefore, to wrong accounts of data characteristics. Researchers should also be aware that the models discussed in the last sections are only two of many possible models that lead to a multiplicative term.

Although these models can differ widely in meaning, they are often treated as equivalent in that the model that involves both main effects and the product term is suggested for testing all interactions (see also Aiken & West, 1991).

10.3   Variable Characteristics

This section addresses three issues that arise when including multiplicative terms in multiple regression:

1. Multicollinearity

2. Leverage points

3. Specific problems concerning ANOVA-type interactions

10.3.1 Multicollinearity of Multiplicative Regression Terms

Table 10.1 indicated that the intercorrelations between the variable $TG *$ $CC1$ and its constituents were very high. Specifically, the correlation between $TG * CC1$ and TG was $r = 0.483$, and the correlation between $TG * CC1$ and $CC1$ was $r = 0.792$. Another example appears in Table 10.3.

In a fashion analogous to Table 10.1, the variable that resulted from element-wise multiplying two other variables with each other, $EDUC *$ $CC1$, is highly correlated with each of its constituents, $EDUC$ and $CC1$.

These examples are indicative of a ubiquitous problem with multiplicative terms in regression analysis: Variables that result from element-wise multiplying two variables with each other tend to be highly correlated with their constituents. As a result, researchers face possibly severe multicollinearity problems. It is easy to imagine that models that involve both $CC1$ and $EDUC * CC1$ can suffer from severe multicollinearity problems. Readers may wish to go through Section 8.2 again for countermeasures to multicollinearity.

A method often used since a paper by Cohen (1978) is termed "partialling regression terms." It involves the following three steps:

1. Calculating multiple regression including only the main effect terms, and not the multiplicative term;

Table 10.3: *Intercorrelations of Cognitive Complexity, CC1, Educational Level, EDUC, the Multiplicative Term, EDUC*CC1, and Recall, REC*

	CC1	EDUC	EDUC * CC1
EDUC	0.301	1.000	
EDUC * CC1	0.941	0.574	1.000
REC	0.119	0.143	0.127

2. Saving residuals from Step 1; and

3. Calculating simple regression only including the multiplicative term.

The reasoning that justifies these three steps is as follows: Step 1 allows one to estimate the contribution made by the predictors. There are no more than the usual multicollinearity problems, because the multiplicative term is not part of the analysis. Step 2 saves that portion of the criterion variance that was not accounted for by the main effect terms. That portion consists of two parts. The first of these is that portion that can be explained by the multiplicative term. The second is that portion that cannot by covered by the complete regression model. Step 3 attempts to cover that portion of criterion variance that the main effects cannot account for. Again, there is no multicollinearity problem, because the predictor main effect terms, highly correlated with the interaction term, are not part of the regression equation.

Cohen's procedure allows one to consider the distinction between regression interaction and the multiplicative or product term in the regression equation. The product term carries the regression interaction. However, because of the possible multicollinearity with the main effect terms it must not be confounded with the regression interaction itself.

It should be noted that residuals are always model-specific. This applies also to residuals as specified by Cohen's procedure. If some other model is specified, residuals from this model should be used to represent the interaction.

10.3.2   Leverage Points as Results of Multiplicative Regression Terms

In addition to multicollinearity, the presence of leverage points is often a problem in the analysis of multiple regression with product terms. Multiplying the elements of two predictors with each other can create the following problems concerning the size of predictors:

1. Scale values are created that cannot be interpreted. Consider the following example. A researcher multiplies the values from two 5-point scales with each other. The resulting products can range from 1 to 25. While numerically correct, values greater than 5 are often hard to interpret. They exceed the range of scale values.

2. Leverage cases are created. This problem goes hand in hand with the first. The larger the products are, the more likely there are leverage cases.

The following numerical example illustrates the problem with leverage cases. The example involves two predictors, X_1 and X_2, and criterion Y. It is assumed that X_2 affects the slope of the regression of Y on X_1. Data are available for five cases. Table 10.4 displays these data, along with estimates, residuals, leverage values, and studentized residuals.

Regressing Y on X_1 under the assumption that X_2 affects the slope parameter yields

$$Y = 2.36 + 1.48X_1 + 0.06 * X_1X_2 + \text{Residual.}$$

Because of high predictor intercorrelation and low power, none of the parameters are statistically significant. However, there are two leverage cases. These are Cases 2 and 5 in Rows 2 and 5 of Table 10.4. In addition, there is one outlier, Case 4.

Figure 10.5 displays the Estimate by Residual plot. Size of data points varies with leverage value. The figure suggests that the two leverage points are located near the regression line. This is to be expected from the definition of leverage points (see Section 5.1). In the present example, the two leverage points are located at the ends of the distribution of estimates, thus illustrating the two problems listed above.

Section 5.2 presents remedial methods to deal with outlier problems.

Table 10.4: *Raw Data and Results of Residual Analysis*

Raw Data				Results of Residual Analysis			
Y	X_1	X_2	X_1X_2	Estimate	Residual	Leverage	Student
7	5	3	15	10.61	−3.61	0.38	−0.51
3	0	2	0	2.36	0.64	0.98	0.53
4	4	1	4	8.50	−4.50	0.31	−0.65
19	5	4	20	10.89	8.11	0.33	9.39
20	7	20	140	20.64	−0.64	0.99	−1.50

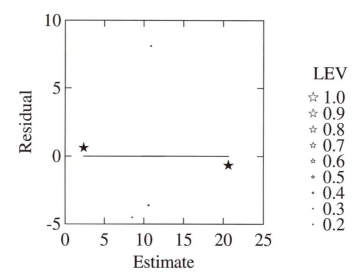

Figure 10.5: Leverage cases in regression with product term.

10.3.3   Problems with ANOVA-Type Interactions

This section deals with specific problems with multiplicative interaction terms that can be best illustrated using ANOVA-like interactions (cf. Rovine & von Eye, 1996). In this type of interaction one does not assume that slope or intercept parameters vary as some monotonic or polynomial function of some other variable. Rather, one assumes that the regression relationship is constant over some range of variable values, is constant again but with different parameters over some other range of variable values, and so on. The ranges are non-overlapping but not necessarily exhaustive; that is, they do not always cover the entire range of observed criterion or predictor values.

Ranges can result from two procedures. One is to categorize a variable. Examples of splits include the median (dichotomization) or the 33 percentile splits. Another way to specify ranges is theory-guided. One can, for instance, define the range of geniuses on the IQ scale, the range of accident-prone car drivers, or the range of binge-drinking alcoholics. For the purposes of this chapter, we define a range as a segment of a variable where we assume the regression relationship to be constant. ANOVA-like regression interactions refer to this type of range.

While the statistical power of detecting regression interaction is generally low, it can be shown (Rovine & von Eye, 1996) that it is, in addition, not equal across the ranges of some variable. Specifically, the power is particularly low when the location of an interaction is in the interior of the product distribution. It is proposed that transforming original predictor values into a set of effect coding vectors (as one would do in ANOVA interaction testing) gives one the best chance of showing an interaction when it indeed exists.

Example

Consider the situation displayed by Table 10.5. This table crosses the two predictors Cognitive Complexity, *CC*, and Educational Level, *EL*. *CC* is split in the three levels of low, medium, and high. *EL* is split in the three levels of no high school diploma, high school diploma, and higher.

Both variables, Cognitive Complexity and Educational Level are predictive of recall performance. The product of *CC* and *EL* forms a bivariate distribution. An ANOVA-like interaction would exist if for some combination of ranges of these two variables a higher (or lower) value on the criterion variable would be measured than one would expect if there were a uniform, monotonic relationship. In the example, consider the range of subjects with high Cognitive Complexity and higher than high school education (lowest right–hand cell in Table 10.5). If for these subjects recall rates are higher (or lower) than one would expect, then there exists an ANOVA-like interaction. An interaction term in regression would have to explain this unique portion of variance.

Table 10.5: *Cross-Classification of the Categorized Variables Cognitive Complexity (CC), and Educational Level*

	Educational Level		
	No high school	High school	Higher
Low CC			
Medium CC			
High CC			x

This interaction (and the one in the top left cell) occurs at the extreme of the bivariate distribution. It could, however, have occurred in other cells as well. Interactions can occur anywhere in the joint distribution of two predictors. If interactions occur in other cells than the extreme ones, they are said to occur within the interior of the product distribution. Table 10.5 displays possible locations.

In any case, one can describe interactions by the portion of variance of a particular combination of predictor values that is not accounted for by the main effect regression terms.

For the following considerations suppose that each of the variables that were crossed to form Table 10.5 has categories with values 1, 2, and 3. Table 10.6 displays the values that the multiplicative term of the two predictors, Educational Level and Cognitive Complexity, assumes for each cell. The values in the cells of Table 10.6 result from multiplying values of variable categories with each other.

Table 10.6 illustrates two important points:

- Element-wise multiplication of predictor values increases the range of values. In the example the range increases from 1 - 3 to 1 - 9. As a result, the variance of the multiplicative variable typically is greater than the variance of its constituents.

- Cases that differ in predictor patterns can be indistinguishable in the value of the product variable. If product terms are created as was done in Table 10.6, cases in cells with inverted indexes have the same product variable value. In the example of Table 10.6, this applies to the cases in cell pairs 12 & 21, 13 & 31, and 23 & 32.

Table 10.6: *Cross-Classification of the Categorized Variables Cognitive Complexity and Educational Level: Values of Product Term*

	Educational Level		
	No high school (1)	High school (2)	Higher (3)
Low CC (1)	1	2	3
Medium CC (2)	2	4	6
High CC (3)	3	6	9

From this second characteristic of product variables a problem results. The problem is redundancy. Cells in the off-diagonal are redundant in the way described. Cells in the diagonal can be unique. The extreme cells at both ends of the diagonal are always unique. Diagonal cells between the extremes are not necessarily unique in their values.[6] Because of uniqueness and leverage, cells with unique values tend to have particular influence on the size of the regression coefficient. Contributions made by interactions in the interior of the product distribution will be relatively smaller, that is, harder to detect.

Centering has often been proposed as a means to reduce problems with the interaction of multiplicative variables. Table 10.7 shows the values of the product term after centering the main effect variables. The values in the cells of Table 10.7 result from performing two steps. The first is centering the main effect variables. The second is creating the multiplicative variable's values by element-wise multiplying the centered main effect variables.

Table 10.7 illustrates two characteristics of multiplicative interaction variables that are created from centered main effect variables:

- The rank order of variable category values changes;

- The redundancies are more severe than before: there are only three different values in the product term; five cells have the value 0.

Performing regression analysis involving simultaneously centered main effect variables and the resulting multiplicative interaction term typically yields the following results:

- The parameter estimates and the significance values for the centered and the noncentered multiplicative interaction variable will be the same;

- The parameters for the centered main effect variables tend to be larger.

---

[6]Readers are invited to demonstrate this using a 4 x 4 Table with variable values ranging from 1 to 4. Only three of the four diagonal cells will have unique values.

Table 10.7: *Cross-Classification of the Categorized Variables Cognitive Complexity and Educational Level: Values of Product Term*

	Educational Level		
	No high school (-1)	High school (0)	Higher (1)
Low CC (-1)	1	0	−1
Medium CC (0)	0	0	0
High CC (1)	−1	0	1

10.3.4   The Partial Interaction Strategy

Repeatedly proposed (Afifi & Clark, 1990; Cronbach, 1987; Rovine & von Eye, 1996), the partial interaction strategy models that part of the residual that is systematically related to the independent variables. The strategy involves the following three steps:

1. Categorization of each variable. This step can be performed in an exploratory and explanatory fashion. The former searches for segments on both the dependent and the independent variable sides that allow one to establish (partial) relationships. This search may involve more than one set of splits. The latter derives splits from theory or earlier results. In either case, researchers should be aware that categorizing typically reduces the power to detect statistically significant relationships: The fewer the segments, the lower the power.

2. Creating a coding variable that identifies cases in cells. The typical coding variable assigns a 1 to each case in the cell assumed to be the location of a partial interaction and a -1 to each case in other cells. Alternatively, specific assumptions can be modeled for particular cells.

3. Estimating regression parameters for the model that involves the residuals of the main effect regression model as criterion variable and the coding variables that specify the partial interactions as predictors.

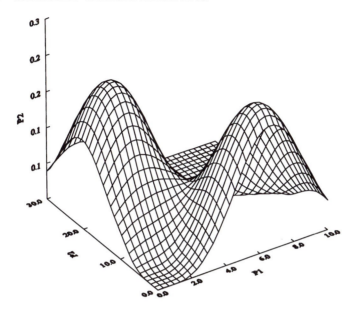

Figure 10.6: Smooth response surface for the artificial data example of partial interaction.

Data Example

The following data example was created to illustrate the partial interaction strategy. The example involves the two predictors, P_1 and P_2, and a criterion, C, measured on eight cases. The analysis of the predictor–criterion relationship involves the following three steps:

Data Description. To obtain a first impression of the predictor-criterion relationship, we create a 3D representation of the data. Rather than plotting the raw scores, we smooth the response surface using the kernel estimator provided in SYSTAT's GRAPH module. The resulting rendering appears in Figure 10.6.

The figure suggests that the criterion has two peaks. The first peak is located where P_1 assumes low values and P_2 assumes high values. This is the left peak in Figure 10.6. The second peak, on the right–hand side in Figure 10.6, is located where P_1 is high and P_2 is low. Between the two peaks, criterion values are small. Raw data appear in Table 10.8.

Estimation of a Main Effect Multiple Regression Model. This model

Table 10.8: *Raw Data for Illustration of Partial Interaction Strategy*

Case	Raw Data					
Number	P_1	P_2	C	I_1	I_2	Residuals
1	1	25	5	1	0	7.35
2	3	22	6	-1	0	−4.75
3	2	18	3	-1	0	−4.90
4	1	17	4	1	0	0.97
5	5	5	33	0	1	−0.26
6	6	4	36	0	-1	−3.47
7	7	2	49	0	-1	2.64
8	8	1	55	0	1	2.43

involves only the two predictors P_1 and P_2. No interaction term is part of this model. For the data in Table 10.8 we estimate the following parameters:

$$C = 5.46 + 0.04 * P_1 - 0.16 * P_2 + \text{Residual}.$$

The linear response surface from this equation appears in Figure 10.7. The residuals from this equation are listed in the last column of Table 10.8.

As is obvious from Figure 10.7, the linear regression model accounts for a large portion of the criterion variance ($R^2 = 0.95$). However, none of the predictors have a significant slope parameter ($t_1 = 1.19, p_1 = 0.29; t_2 = -2.02, p_2 = 0.10$). The residuals from this analysis appear in the last column of 10.8.

Specification of Partial Interaction Terms. The data were constructed so that there is a strong negative correlation between P_1 and P_2,[7] and that P_1 and P_2 each explains a large portion of the criterion variance. In addition, the data show nonlinear variation for values high in P_1 and low in P_2, and for values low in P_1 and high in P_2. The cutoff is after the fourth case in Table 10.8. Therefore, after extracting that portion of the criterion

---

[7] Readers are invited to determine whether there are multicollinearity problems in these data.

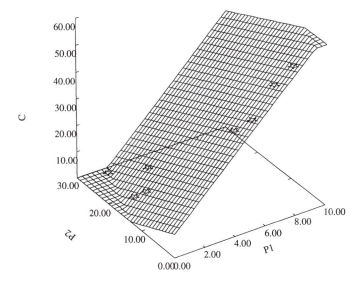

Figure 10.7: Multiple linear regression with no interaction terms for the data in Table 10.8.

variance that can be accounted for by the main effects of P_1 and P_2, we specified two interaction vectors, I_1 and I_2, each designed to explain nonlinear variation in the residuals. The first vector, I_1, was designed to explain nonlinear variance for the P_1-low and P_2-high sector. The second vector, I_2, was designed to explain nonlinear variance in the P_1-high and P_2-low sector. For the sake of simplicity, we designed both interaction vectors so that they represent quadratic trends. Both interaction vectors appear in Table 10.8. Regressing the residuals from the main effect model onto I_1 and I_2 yields the following equation:

$$\text{Residual} = 0.00 + 4.49 * I_1 + 0.75 * I_2 + \text{Residual}_2.$$

The two interaction vectors allow one to explain the $R^2 = 0.66$ of the variance of the residuals from the main effect model. Only the first of the interaction vectors has a significant regression slope ($t = 3.05, p = 0.03$). The second has a slope not different than zero ($t = 0.51, p = 0.63$).

Caveat

This example illustrates that custom-tailored specification of regression interaction terms allows one to explain substantial portions of variance that cannot be explained from the multiple regression main effect model. It is important to realize, however, that without guidance by theory, the search for interaction sectors can carry researchers to data fitting, that is, description of sample-specific data characteristics. Such characteristics are often unique to samples. Therefore, unless theory dictates where and what type of interaction to expect, replications are strongly recommended before publishing results.

Chapter 11

ROBUST REGRESSION

This chapter introduces readers to the concepts of robust regression. Specifically, after a brief description of the concept of robustness (Section 11.1) and after presenting ridge regression (Section 11.2.1), Least Median of Squares (LMS) regression (Rousseeuw, 1984) (Section 11.2.2), and Least Trimmed Squares (LTS) regression (Section 11.2.3), we briefly describe M-estimators. Section 11.3 covers computational issues.

11.1 The Concept of Robustness

Robustness can be defined as "insensitivity to underlying assumptions – for example, about the shape of the distribution of measurements" (Hoaglin, Mosteller, & Tukey, 1983, p. 283). Similarly, Hartigan (1983, p. 119) calls a statistical procedure robust "if its behavior is not very sensitive to the assumptions which justify it."

Most investigations of robustness focus on parametric or distributional assumptions. These are assumptions about a probability model for the observations under study and about a loss function connecting statistical decision and an unknown parameter value.[1] Accordingly, a large portion of robustness investigations examine the robustness of statistical tests. A statistical test is considered robust if its "significance level ... and

---

[1] Notice that in Bayesian statistics an additional prior distribution needs to be considered.

power ... are insensitive to departures from the assumptions on which it is derived" (Ito, 1980, p. 199). Results of such investigations suggest, for example, that the F test used in regression analysis and analysis of variance is remarkably robust against heterogeneity of variance and non-normality, in particular when group sizes are equal, that is, in balanced designs. When groups differ in size and variances differ across groups, robustness is clearly less pronounced (Ito, 1980).

In a similar fashion, there have been investigations of the effects of other types of violations of assumptions. For example, von Eye (1983) performed a simulation study on the effects of autocorrelation on the performance of the t test. Results suggested that positive autocorrelations lead to inflated values of the t statistic, and negative autocorrelations lead to deflated values of the t statistic. These biases increase with the magnitude of correlation.

More recent examples of simulation studies concerning the robustness of estimators include the work by de Shon and Alexander (1996). The authors examined the following six tests of regression slope homogeneity: the F test, the χ^2 test, James's test, the normalized t test, the Welch-Aspin F^* test, and Alexander's normalized t approximation. Various violations of the conditions for proper application of the standard F test were simulated, for instance, nonnormality of the dependent variable, Y, in one or both populations, heterogeneity of error variances, and nonorthogonal designs. Results suggest that none of the tests perform well under all conditions. However, when the ratios of Y variance to X variance are approximately equal, the χ^2 test seems to perform well. It has more power than the F test or any of the approximations. When the ratio of the largest group error variance to the smallest group error variance is smaller than 1.50, Alexander's normalized t approximation should be used (Alexander & Govern, 1994). Ito (1980) notes that, in general, robustness cannot be exhaustively investigated because there are more ways to violate assumptions than to satisfy them. Yet, it is important to know effects of the most frequent and most important violations.

One very important aspect of robustness has to do with outliers, specifically, leverage outliers. The question that arises in this context concerns the extent to which estimation of regression parameters is affected by such outliers. It is well known that the OLS method is sensitive to such outliers. Consider the following example. A newscaster tries to answer the question whether the number of years a shooting guard has spent in the National Basketball Association league allows one to predict the average number of points per game scored. The newscaster draws a random sample of $n = 13$ guards, at different points in their careers, from different cohorts. Table 11.1 shows the number of years a guard had spent in the league, and the average number of points scored in the last of these years.

Regressing the number of points scored onto the number of years yields the following regression equation:

$$\text{Points} = 7.61 + 0.67 * \text{Year} + \text{Residual}.$$

The slope parameter estimate of this equation is not significant ($t = 1.03; p = 0.33$). Figure 11.1 shows the raw data and the thick regression line.

The graph in Figure 11.1 suggests that the thick line is a poor representative of the relationship between career length and scoring performance. The main reason is that there is an outlier. One player, nine years in the league, had a scoring average of 29 points per game, far above the crowd.

The thin regression line depicts the relationship between number of years in the league and points scored, under exclusion of the outlier. In contrast to the thick line, the slope of the thin line is negative, the regression equation being

$$\text{Points} = 11.61 - 0.55 * \text{Years} + \text{Residual}.$$

In addition, the relationship now is statistically significant ($t = -2.46$; $p = 0.03$).

The comparison of these two results illustrates the effects an outlier can have on an OLS estimate of a regression slope. Estimators are robust

Table 11.1: *Number of Years Spent in the Basketball League and Average Number of Points Scored in Last Year*

Year	2	3	5	6	6	7	7	3	2	1	9	4	4
Pts	12	7	12	9	10	6	6	9	10	11	29	9	10

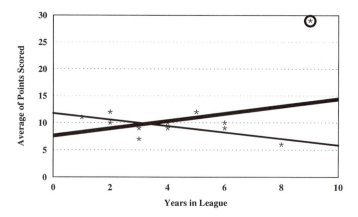

Figure 11.1: Regression of number of points scored on number of years in the league before (thick line) and after (thin line) elimination of outlier.

if they are insensitive to the presence of outliers or extreme values. This chapter focuses on approaches to creating robust estimates of regression slopes. The next section reviews sample models of robust linear regression. It includes the Least Median of Squares and the Least Trimmed Squared approaches in more detail, and a brief introduction to M-estimators.

11.2 Models of Robust Regression

This section provides a brief review of models for robust regression. Models are reviewed in two groups. The first includes ridge regression. The second includes so-called M-estimators for regression parameters (Goodall, 1983; Hettmansperger & Sheather, 1992; Marazzi, 1980).

11.2.1 Ridge Regression

The following approach of ridge regression is typically discussed in the context of solving multicollinearity problems (e.g., Afifi & Clark, 1990). However, estimators from ridge regression tend to be robust. Therefore, we review ridge regression in the present context (see also Fennessey & D'Amico, 1980; Gruber, 1989).

Consider the situation in which there are two estimates of some parameter. Whereas the first is unbiased, the second has a slight bias but is considerably more precise than the first. In a situation like this, many researchers will prefer the slightly biased but more precise estimate, because it has a higher probability of being close to the true estimator. Ridge regression is a method for estimating such biased but precise estimators of slope and intercept.

Ridge regression starts from the standard regression model with an estimated parameter vector

$$\mathbf{b} = (\mathbf{X'X})^{-1}\mathbf{X'y}. \tag{11.1}$$

Now, assume that all variables are standardized, that is,

$$Z_j = \frac{X_j - \bar{X}_j}{\sigma_j},$$

where j indexes variables. Then, (11.1) can equivalently be expressed as

$$\mathbf{b} = \mathbf{R_{xx}^{-1}R_{yx}},$$

where $\mathbf{R_{xx}}$ is the correlation matrix of the x variables, that is, the regression predictors, and $\mathbf{R_{yx}}$ is the vector of the correlations of the predictors with the criterion, y. More specifically, we have

$$\mathbf{R_{xx}} = \begin{pmatrix} 1 & r_{12} & r_{13} & \cdots & r_{1p} \\ r_{21} & 1 & r_{23} & \cdots & r_{2p} \\ \vdots & \vdots & \vdots & \ddots & \vdots \\ r_{p1} & r_{p2} & r_{p3} & \cdots & 1 \end{pmatrix},$$

with $r_{ij} = r_{ji}$, for the predictors x predictors correlation matrix, and

$$\mathbf{R_{yx}} = \begin{pmatrix} r_{1y} \\ r_{2y} \\ \vdots \\ r_{p1} \end{pmatrix},$$

where $j = 1, \ldots, p$ indexes predictors.

To estimate parameters for ridge regression from standardized variables, one introduces a biasing constant $c \geq 0$ as

$$\mathbf{b^R} = (\mathbf{R_{xx}^R})^{-1}\mathbf{R_{yx}}$$

with

$$\mathbf{R_{xx}^R} = \mathbf{R_{xx}} + \mathbf{cI},$$

or, more specifically,

$$\mathbf{R_{xx}^R} = \begin{pmatrix} 1+c & r_{12} & r_{13} & \cdots & r_{1p} \\ r_{21} & 1+c & r_{23} & \cdots & r_{2p} \\ \vdots & \vdots & \vdots & \ddots & \vdots \\ r_{p1} & r_{p2} & r_{p3} & \cdots & 1+c \end{pmatrix}. \tag{11.2}$$

The constant c represents the magnitude of bias introduced in ridge regression. When $c = 0$, the solution is OLS. When $c > 0$, there is a bias. However, solutions from $c > 0$ are usually more stable, that is, robust, than OLS solutions.

From an applied perspective most important is the determination of a value for the constant, c. The optimum value of c may be specific to data sets and, therefore, needs to be found for each data set. One method to find the optimum value of c uses information from two sources: the Variance Inflation Factor (VIF; see Section 8.1) and the ridge trace. The ridge trace is the curve of the values assumed by a regression slope estimate when c increases. Typically, after fluctuations, the estimate changes its value only slightly when c further increases. Simultaneously, the VIF falls rapidly as c first moves away from 0, and changes only slightly when c increases further. The value of c finally chosen balances (1) the VIF, which must be sufficiently small, (2) the slope coefficient(s) which must change only minimally when c is increased further, and (3) c itself, which, in order to avoid excessive bias, should be as small as possible.

Alternatively or in addition to the VIF, one inspects the mean squared error of the biased estimator, which is defined as the variance of the estimator plus the square of the bias.

The following two artificial data examples illustrate the use of the ridge trace (Afifi & Clark, 1990, pp. 240ff). Consider a criterion, Y, that is predicted from two independent variables, X_1 and X_2. Suppose all three variables are standardized. Then, the OLS estimator of the standardized regression coefficient for X_1 is

$$b_1 = \frac{r_{1y} - r_{12}r_{2y}}{1 - r_{12}^2}, \qquad (11.3)$$

and the OLS estimator of the standardized regression coefficient for X_2 is

$$b_2 = \frac{r_{2y} - r_{12}r_{1y}}{1 - r_{12}^2}. \qquad (11.4)$$

The ridge estimators for these two regression coefficients are

$$b_1^R = \frac{r_{1y} - \frac{r_{12}}{1+c}r_{2y}}{1 - \left(\frac{r_{12}}{1+c}\right)^2 (1+c)} \qquad (11.5)$$

and

$$b_2^R = \frac{r_{2y} - \frac{r_{12}}{1+c}r_{1y}}{1 - \left(\frac{r_{12}}{1+c}\right)^2 (1+c)}. \qquad (11.6)$$

For the next numerical example we proceed as follows: First we specify values for the three variable intercorrelations. Second, we insert into (11.5) and (11.6) varying values of c. For the variable intercorrelations we select the following values: $r_{12} = 0.8, r_{1y} = 0.5$, and $r_{2y} = 0.1$. For c we specify the range of $0 \le c \le 8$ and an increment of 0.1. Thus, we run 81 iterations. For each iteration we determine the values of the two regression coefficients, b_1 and b_2. The curve that depicts these values is the *ridge trace*. Figure 11.2 displays the ridge trace for the present correlations for the first 50 iterations.

Before discussing the figure we insert the correlations into (11.3) and (11.4). We obtain the standard OLS estimates $b_1 = 1.17$ and $b_2 = -0.83$. These are also the first results in the iteration ($c = 0$).

The figure suggests that with each increase of c the values of the regression coefficients change. Specifically, b_2 decreases and b_1 increases. In

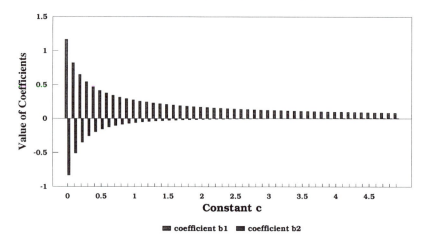

Figure 11.2: Ridge traces for artificial data example.

this example, coefficient b_2 assumes positive values after the 30th iteration. It keeps increasing until the 68th iteration. After that, it decreases again and both coefficients asymptotically approach zero.

When selecting a ridge constant one selects that value after which there is no longer any substantial change in the regression coefficients. In the present example, this seems to be the case after $c = 0.7$. Notice that, even if the VIF and the mean squared error are also used for selection of c, the selection is still a matter of subjective decision making, because there are no objective criteria that could be used to guide decisions. As a rule of thumb, one can say that values of c between 0 and 1 are often the most interesting and promising ones.

Figure 11.2 presents a graph that is typical of ridge traces. The shape of traces, however, depends only on the pattern of variable intercorrelations. This is exemplified in a second numerical example. This example uses the following correlations: $r_{12} = 0.30, r_{1y} = 0.25$, and $r_{2y} = 0.10$. The standardized regression coefficients for these correlations are $b_1 = 0.24$ and $b_2 = 0.03$. Figure 11.3 displays the ridge trace for the first 50 values of c, beginning with $c = 0$ and using an increment of 0.1.

As in the first example, the two regression coefficients approach zero as c increases. After about $c = 0.6$, there are no substantial changes in the coefficients when considering small to moderate changes in c. Thus,

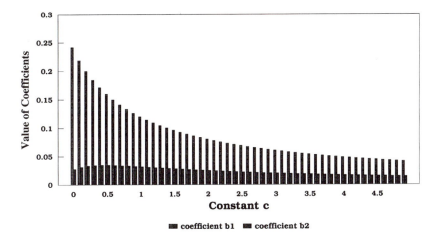

Figure 11.3: Ridge traces for two positive slope coefficients.

we may select $c = 0.6$ as the ridge constant for this example.

The following data example investigates two predictors, Breadth of Cognitive Complexity ($CC1$) and Depth of Cognitive Complexity ($CC2$), and a criterion, Text Recall (REC). In a sample of $n = 66$ adult females, these three variables correlated as shown in Table 11.2.[2].

Table 11.2: *Intercorrelations of the Predictors, CC1 and CC2, and the criterion, REC*

	CC1	CC2
CC2	0.066	
REC	-0.074	0.136

Table 11.2 suggests that the variable intercorrelations are low. Thus, we cannot expect to explain large portions of the criterion variance. The highest correlation is the one between the two predictors. Standard OLS

---

[2]The following calculations are performed using the correlation matrix method described in Section 11.3.1. This method estimates the slope parameter from a correlation matrix. Correlations do not contain information on the means of variables. Therefore, regression models from correlation matrices do not include the intercept parameter.

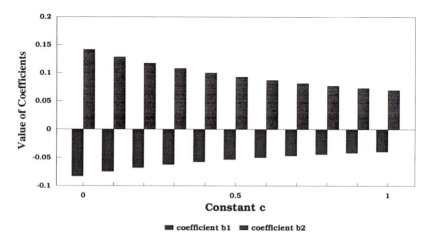

Figure 11.4: Ridge traces for cognitive data.

regression of *REC* onto *CC1* and *CC2* explains no more than 2.5% of the variance, and the regression model fails to be statistically significant ($F_{2,64} = 0.83; p > 0.05$). Accordingly, neither regression slope is significantly different than zero. The following OLS regression equation was calculated[3]:

$$REC = -0.08 * CC1 + 0.14 * CC2 + \text{Residual}.$$

To perform a ridge regression we first create a ridge trace. In this example, we use values for the constant that vary between 0 and 1 in steps of 0.1. Figure 11.4 displays the resulting ridge trace.

In addition to the ridge trace we consider the standard error of the estimates. These estimates are the same for the two predictors. Therefore, only one needs to be graphically displayed. Figure 11.5 contains the standard error for the 11 ridge regression runs.

---

[3]Notice that in the present context we estimate regression slopes that go through the origin (no intercept parameter is estimated). This procedure is put in context in the section on computational issues (Section 11.3).

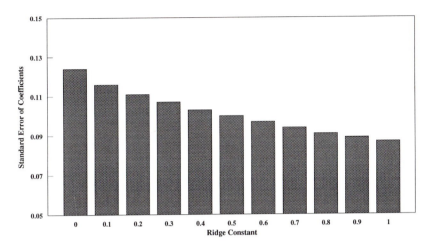

Figure 11.5: Change in standard error of regression estimates in cognition data.

Figure 11.4 suggests that values greater than $c = 0.4$ do not lead to a substantial decrease in parameter estimates. Figure 11.5 shows an almost linear decrease in standard error with slightly bigger decreases for smaller c values than for larger c values. From these two sources we select a ridge constant of $c = 0.4$. The regression coefficients for this solution are $b^R_{CC1} = -0.057$ and $b^R_{CC2} = 0.100$. Both coefficients are smaller than the original OLS coefficients for which $c = 0$.

Problems with Ridge Regression

The following paragraphs review three problems with ridge regression. The first is the bias that comes with ridge regression solutions with $c > 0$. It is one of the virtues of ordinary least squares that its solutions are unbiased. Unless there are strong reasons why a (slightly) biased solution is preferable, this advantage should not be abandoned.

Second, it has to be emphasized again that the decision for a constant, c, is a subjective one. There are still no methods for objectively determining a constant from a sample. The main argument supporting ridge regression is that ridge estimates will perform better than OLS estimates in the population or, more specifically, will provide better predictions in the population. However, ridge regression estimates do not explain as much of the sample criterion variance as do OLS estimates.

This can be illustrated using the previous example. As was indicated before, the standard OLS solution with *CC1* and *CC2* as predictors and *REC* as criterion has an R^2 of 0.025. The ridge regression solution only has an R^2 of 0.018. With increasing c the portion of variance accounted for decreases even more. For instance, when $c = 1$, we calculate $R^2 = 0.012$.

Third, there are no statistical significance tests for solutions from ridge regression. The tests printed out when simulating ridge regression using standard OLS regression programs must not be trusted (one reason why is explained in Section 11.3 on computational issues).

11.2.2   Least Median of Squares Regression

Introduced by Rousseeuw (1984), the method of Least Median Squares (LMS) allows one to estimate most robust estimates of regression parameters (see also Rousseeuw & Leroy, 1987). As the median, the LMS method has a breakdown point of 50%. That is, up to 50% of the data can be corrupt before parameter estimates are substantially affected.

To introduce readers to LMS regression we draw on the concepts of OLS regression. We start from the regression model

$$y = X\beta + \epsilon,$$

where $\mathbf{y}$ is the vector of observed values, $\mathbf{X}$ is the matrix of predictor values (the design matrix), β is the vector of parameters, and ϵ is the residual vector. To obtain an estimate $\mathbf{b}$ of β, OLS minimizes the sum of squared residuals,

$$(\mathbf{y} - \mathbf{Xb})'(\mathbf{y} - \mathbf{Xb}) \longrightarrow min. \qquad (11.7)$$

or, in other terms,

$$\sum_i e_i^2 \longrightarrow min, \qquad (11.8)$$

In contrast, Rousseeuw (1984) proposes minimizing the median of the squared residuals rather than their sum,

$$md(e_i^2) \longrightarrow min,$$

where $md(e_i^2)$ is the median of the squared residuals.

Solutions for (11.7) are well known (see Section 3.1). These are closed-form solutions, that is, they can be calculated by the one-time application of a set of formulas. In contrast, there is no closed form that can be used to solve (11.8). Therefore, iterative procedures have been devised that typically proceed as follows (Rousseeuw & Leroy, 1987):

1. Select a subsample of size n_j and calculate the median of the squared residuals along with the regression parameter estimates; save median and parameter estimates;

2. Repeat step (1) until smallest median has been found.

The following example uses data from the statistical software package S-Plus (Venables & Ripley, 1994). The data were first published by Atkinson (1986). The data describe the three variables Distance, Climb, and Time from Scottish hill races. Distance is the length of a race, Climb is the height difference that must be covered in a race (in feet), and Time is the record time for a race.

In this example we regress Time on Climb. In order to be able to compare LMS regression with OLS regression we first perform a standard OLS regression. We obtain the regression equation

$$\text{Time} = 12.70 + 0.025 * \text{Climb} + \text{Residual}.$$

The slope estimate is significant ($t = 7.80; p < 0.01$). Figure 11.6 displays the raw data and the OLS regression line.

The figure suggests that, although there clearly is a trend that higher climbs require longer times to complete the race, there are exceptions. Some of the shorter races seem to take longer to complete than one would predict from the regression relationship. In particular, one of the races that covers a climb of about 2000 feet is far more time consuming than one would expect.

One may suspect that this data point (and a few others) exert undue leverage on the regression slope. Therefore, we perform, in a second run,

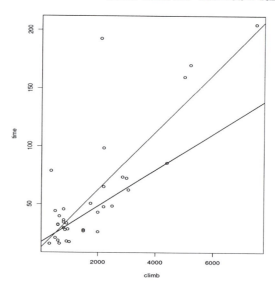

Figure 11.6: OLS and two robust methods of regression solutions for hill data.

LMS regression. The slope estimated for LMS regression should not be affected by the presence of outliers. Up to 50% of the data points may be outliers before the LMS regression slope is affected.

The LMS regression equation is

$$\text{Time} = 16.863 + 0.015 * \text{Climb} + \text{Residual}.$$

Obviously, the slope of the LMS regression line is less steep than the slope of the OLS regression line. There is no significance test for LMS regression. However, the program (see Chapter 5) identifies outliers. Six of the 35 race track data points are outliers. Figure 11.6 also presents the LMS regression line. This line is, in the present example, virtually identical to the LTS regression line (to be explained in the next section).

Recently, LMS regression has met with criticism (Hettmansperger & Sheather, 1992). The reason for this criticism is that whereas LMS regression is clearly one of the most robust methods available when it comes to not being affected by bad outliers, it seems to be overly sensitive to bad inliers. These are data points that lie within the data cloud. They look unsuspicious and are hard to diagnose as problematic data points. Yet, they have the leverage to change the slope of LMS regression. Hettmansperger and Sheather (1992) present an artificial data set in which moving one of the bad inlier data points by a very small amount results in a change of the LMS regression slope by 90°.

11.2.3 Least Trimmed Squares

In response to these criticisms, Venables and Ripley (1994) and Rousseeuw and Leroy (1987) recommend using Least Trimmed Squares (LTS) regression. This approach estimates parameters after trimming, that is, excluding observations at both tails of the distribution. The number of observations to be excluded is controlled by choosing q in the following formula,

$$\sum_{i=1}^{q} (y_i - x_i b)^2 \longrightarrow min.$$

LTS regression is more efficient than LMS regression. In addition, it has the same extreme resistance, that is, high breakdown point. There has been a discussion as to what value is best specified for q, that is, the number of observations to be included. Most frequently, one finds the following two definitions:

$$q = \frac{n}{2} + \frac{p+1}{2}$$

and

$$q = \frac{n}{2} + 1.$$

In the examples in this book we use the first definition of q. The regression line estimated using LTS regression is

$$\text{Time} = 16.863 + 0.015 * \text{Climb} + \text{Residual}.$$

As for the LMS regression, there is no significance test except for jackknife procedures.

Figure 11.6 also displays the LTS regression line. Both, the formula

and the figure suggest that the LMS and the LTS regression estimates
are very close to each other. However, it still needs to be investigated in
detail whether LTS regression is generally better that LMS regression, in
particular, in regard to bad inliers.

11.2.4   M-Estimators

When discussing M-estimators of location, Goodall (1983) defines the
M-estimate $T_n(x_1, \dots, x_n)$ for t, given some function $\rho(x; t)$ and sample
$x_1, \dots, x_n$, as that value of t that minimizes the objective function

$$\sum_{i=1}^{n} \rho(x_i; t),$$

where the x_i are the observed values and t is the location estimate. The
characteristics of ρ determine the properties of the estimator. Let the
first derivative of the objective function be called ψ. Then, the minimum
of the function is

$$\sum_{i=1}^{n} \psi(x_i; t) = 0. \qquad (11.9)$$

The best known M-estimate is the sample mean. For this estimate of
location, estimated by least squares, ρ is the square of the residual,

$$\rho(x; t) = (x - t)^2.$$

The expression

$$t = \frac{1}{n} \sum_{i=1}^{n} x_i$$

gives the minimum of (11.9). It is the sample mean.

The ψ function of a robust M-estimator has a number of very desirable properties (Goodall, 1983), the most important of which is that the
breakdown point of the estimator is very large. Here, the breakdown point
of an estimator is the smallest portion of the data that can be replaced
by arbitrary values to cause the estimator to move an arbitrary amount.

Thus, the breakdown point is a measure of an estimator's robustness. The higher the breakdown point, the more robust is an estimator. The highest possible breakdown point is 50%. If more than 50% of the data are contaminated, one may wonder whether the remaining portion that is less than 50% is the contaminated one. The median is an M-estimator with a breakdown point of 50% for the location problem.

The approach of Marazzi (1980, 1993) to describing M-estimators in regression is based on the standard regression model,

$$y = X\beta + \epsilon,$$

where X is the design matrix and the residuals are normally distributed with expectancy 0 and variance σ^2. Marazzi solves the following system of equations for β and σ:

$$\sum_{i=1}^{n} \psi\left(\frac{r_i}{\sigma w_i}\right) w_i x_{ij} = 0, \quad \text{for } j = 1, \ldots, p$$

$$\sum_{i=1}^{n} \chi\left(\frac{r_i}{\sigma w_i}\right) w_i^2 = \text{constant}$$

where r_i is defined as

$$r_i = y_i - x_i'\beta,$$

where x_i is the ith column of the design matrix X. χ and ψ are user-defined functions, the w_i are given weights, and the constant is a given number. Depending on what values are chosen for function parameters, special cases result. For example, for constant = 1 one obtains the approach of Huber (1981).

11.3   Computational Issues

This chapter gives two computational examples of robust regression. The first example applies ridge regression. For the illustration of ridge regression we use SYSTAT for Windows, Release 5.02. The second example illustrates M-estimators, specifically, LMS and LTS regression. It uses the statistical software package S-Plus (Venables & Ripley, 1994).

11.3.1   Ridge Regression

SYSTAT does not include a module for ridge regression. Therefore, we present two equivalent ways to estimate parameters in the absence of such a module. The first involves translating the procedures outlined by Afifi and Clark (1990) for the BMDP statistical software package[4] into the SYSTAT environment. We term this method the Dummy Observation Method. The second option works with the correlation matrix of predictors and criteria. We term it the Correlation Matrix Method.

Afifi and Clark's Dummy Observation Method

The Dummy Observation Method can be employed for estimating ridge regression parameters using practically any computer program that performs ordinary least squares regression. The only requirement is that the program use raw data. To introduce the method consider a multiple regression problem with $p > 1$ predictors, $X_1, X_2, \ldots, X_p$, and criterion Y. The Dummy Observation Method involves the following two steps:

1. Standardization of All Variables. As a result, all variables have a mean of 0 and a standard deviation of 1; no intercept needs to be estimated.

2. Addition of Dummy Observations. For each of the p predictors, one dummy observation is appended to the standardized raw data. These observations are specified as follows: $Y = 0$, and the jth predictor assumes the value

$$X_k = \begin{cases} 0 & \text{if } k \neq j \\ \sqrt{c(n-1)} & \text{if } k = j \end{cases}$$

where c is the ridge constant.

In other words, the appended dummy observations are all zero for Y. For the p predictors, one appends a $p \times p$ matrix with $(c(n-1))^{1/2}$ in the main diagonal and 0 in the off-diagonal cells.

---

[4] A number of statistical software packages do provide modules for ridge regression. Examples include the BMDP4R and the SAS-RIDGEREG modules.

For an illustration consider a data set with the two predictors X_1 and X_2, criterion Y, and n subjects. After appending the p dummy observations, the regression equation appears as follows:

$$
\begin{pmatrix}
y_1 \\
y_2 \\
\vdots \\
y_n \\
0 \\
0
\end{pmatrix}
=
\begin{pmatrix}
1 & x_{11} & x_{12} \\
1 & x_{21} & x_{22} \\
\vdots & \vdots & \vdots \\
1 & x_{n1} & x_{n2} \\
1 & \sqrt{c(n-1)} & 0 \\
1 & 0 & \sqrt{c(n-1)}
\end{pmatrix}
\begin{pmatrix}
b_0 \\
b_1 \\
b_2
\end{pmatrix}
+
\begin{pmatrix}
e_1 \\
e_2 \\
\vdots \\
e_n \\
e_{n+1} \\
e_{n+2}
\end{pmatrix}.
$$

Obviously, this equation involves p observations more than the original regression equation. Thus, estimating regression parameters from this approach artificially inflates the sample size by p. This is one reason why significance tests for ridge regression from this approach must not be trusted. Another reason is that appending dummy observations leads to predictors and criteria with nonmeasured values. Including these values distorts estimates of the portion of variance accounted for. The estimates from this approach are automatically the ridge regression parameter estimates. A ridge trace can be created by iterating through a series of constants, c (see Figure 11.4).

The following sections illustrate this variant of ridge regression using the MGLH regression module in SYSTAT. We use the cognitive complexity data already employed for the example in Figure 11.4. The illustration involves three steps. In the first we append the dummy observations. In the second step we estimate regression slope parameters. In the third step we iterate using a series of constants, c. For the first step we issue the following commands. The sample size is $n = 66$. The value to be inserted in the main diagonal of the dummy observation predictor matrix is $(c(66 - 1))^{1/2}$. For $c = 0.1$ this value is 2.55; for $c = 1$, this value is 8.062. Since we have two predictors, we have to append two dummy observations.

Command	Effect
Use Memsort (CC1, CC2, REC)	Reads variables CC1, CC2, and REC from file "Memsort.sys"; SYSTAT presents list of variables on screen
Click Window, Worksheet	Raw data are pulled into a window on the screen
Click Editor, Standardize	Prepares standardization of variables
Highlight CC1, CC2, and REC; click Add each time a variable is highlighted	Specifies variables to be standardized
Click Ok	SYSTAT responds by asking for the name for a file where it saves the standardized variables
Type "CCRECS"	Specifies file name; data will be saved in file "CCRECS.SYS"
Click Ok	Standardizes and saves selected variables
Back in the Worksheet, hit the END key	Carries us one field past the last data entry
Enter "0" for REC, "2.55" for CC1, and "0" for CC2	Specifies values for the first dummy observation
Enter "0" for REC, "0" for CC1, and "2.55" for CC2	Specifies values for the second dummy observation
Click File, Save	Saves data with appended dummy observations in file "CCRECS.SYS"
Click File, Close	Concludes data editing; carries us back to the SYSTAT command mode window

After these operations we have a data file that is different than a conventional raw data file in two respects. First, the variables that we use in the ridge regression runs are standardized. Second, there are $p = 2$ appended dummy observations that are needed in order to obtain ridge regression parameter estimates from employing the standard OLS regression program. The dummy observations must not be part of the standardization. Therefore, standardization must be performed before appending dummy observations.

Using this new data set we now can estimate a first set of ridge regression parameters. We do this with the following commands:

Command	Effect
Use ccrecs	Reads file "CCRECS.SYS"
Click Stats, MGLH, Regression	Initiates the OLS regression module
Highlight CC1 and CC2 and assign them to Independent; highlight REC and assign it to Dependent	Specifies predictors and criterion for regression
Click Include Constant	Results in Regression Model with no constant (disinvokes inclusion of constant which is default)
Click Ok	Concludes model specification, performs estimation, and presents results on screen

The following output displays the results for this run:

```
SELECT (sex= 2) AND (age> 25)
>MODEL REC = CC1+CC2
>ESTIMATE
Model contains no constant

Dep Var: REC
N: 67
Multiple R: 0.127
Squared multiple R: 0.016

Adjusted squared multiple R: 0.001
Standard error of estimate: 1.050
```

Effect	Coefficient	Std Error	t	P(2 Tail)
CC1	0.062	0.125	0.490	0.626
CC2	0.101	0.120	0.839	0.405

The output suggests that the sample size is $n = 67$. We know that this is incorrect, because we had added two dummy observations. The correct sample size is 65. Therefore, with the exception of the slope parameter estimates, none of the statistical measures and tests provided in the output can be trusted. The regression equation is

$$REC = 0.062 * CC1 + 0.101 * CC2 + Residual.$$

In order to create information for the ridge trace, we repeat this run for a total of 10 times, with increasing c. Table 11.3 contains the values that must be inserted as values for the dummy observations for $0 < c \leq 1$ in increments of $c = 0.1$ and a sample size of $n = 65$.

The ridge trace for this example appears in Figure 11.4.

The Correlation Matrix Method

The Correlation Matrix Method uses the correlation matrix given in Table 11.2, that is, the correlation matrix that includes the biasing constant, c, as the summand for the diagonal elements of the predictor intercorrelation matrix (see Equation (11.2)). In analogy to the Dummy Observation Method, the Correlation Matrix Method performs the following two steps:

1. Calculation of correlation matrix of all variables that includes the predictor variables and the criterion variables (see below)

2. Addition of ridge constant to each of the diagonal elements of R.

Table 11.3: *Values for Dummy Observations for $n = 65$ and $0 < c \leq 1$*

c	0.1	0.2	0.3	0.4	0.5	0.6	0.7	0.8	0.9	1.0
$(65c)^{1/2}$	2.5	3.6	4.4	5.1	5.7	6.2	6.7	7.2	7.6	8.0

Unlike in the Dummy Observation Method, the ridge constant is added as is. No transformation is necessary.

The following sections illustrate this variant of ridge regression using the CORRelation and the MGLH modules in SYSTAT. We use the same data as in the last section, that is, the data that relate two measures of cognitive complexity, *CC1* and *CC2*, to prose recall, *REC*, in a sample of $n = 65$ adult females. The illustration involves the following steps: First, we estimate the predictor x predictor correlation matrix and the predictors x criterion intercorrelation vector. Second, we estimate parameters for OLS and ridge regression.

First, we create and store the matrix of variable intercorrelations. We create the complete correlation matrix of the variables, *CC1*, *CC2*, and *REC*, because the last row of the lower triangular of this matrix contains the vector of predictor–criterion intercorrelations. The following commands yield the correlation matrix:

Command	Effect
Use Memsort (CC1, CC2, REC)	Reads variables CC1, CC2, and REC from file "Memsort.sys"; SYSTAT presents list of variables on screen
Click Stats, Correlation, Pearson	Selects the Pearson correlation from the correlation module
Highlight all three variables, click Add	Includes all three variables in computation of correlation matrix
Click the Save File square and Ok	Indicates that correlation matrix is to be stored; SYSTAT requests name of file
Type "CORRCC"	Specifies file name for correlation matrix
Click Save	Performs correlations; saves correlation matrix in file with double precision; presents correlation matrix on screen; and indicates that correlation matrix has been saved

The following output reproduces the result of these operations:

```
PEARSON CC1 CC2 REC

Pearson correlation matrix

              CC1             CC2             REC
CC1         1.000
CC2         0.132           1.000
REC         0.077           0.120           1.000

Number of observations: 65
```

To make sure the same results can be achieved from the correlation matrix as from the raw data, we issue the following commands:

Command	Effect
Use Corrcc	Reads triangular file "Corrcc.sys" which contains correlation matrix of variables CC1, CC2, and REC
Click Stats, MGLH, and Regression	Opens window for specification of regression model
Assign CC1 and CC2 to Independent and REC to Dependent	Specifies which variables are independent and which is dependent
Type "66" in number of cases rectangle	Specifies number of cases (required when analyzing correlation matrices)
Click Ok	Starts regression run and presents results on screen; notice that the program automatically calculates a model with no intercept when processing correlation matrices

The following output reproduces the result of this regression run:

```
REGRESS
>MODEL REC = CC1+CC2 /N=65
>ESTIMATE
```

Dep Var: REC
N: 65
Multiple R: 0.135
Squared multiple R: 0.018

Adjusted squared multiple R: 0.0
Standard error of estimate: 0.999

Effect	Coefficient	Std Error	t	P(2 Tail)
CC1	0.062	0.127	0.488	0.628
CC2	0.112	0.127	0.880	0.382

Analysis of Variance

Source	SS	df	MS	F	P
Regression	1.144	2	0.572	0.573	0.567
Residual	61.856	62	0.998		

For reasons of comparison we also present the output from a standard regression run, that is, a run that uses raw data and includes an intercept term.

Results from this run appear in the following output.

```
SELECT (SEX= 2) AND (age> 25)
>MODEL REC = CONSTANT+CC1+CC2
>ESTIMATE
```

Dep Var: REC
N: 65
Multiple R: 0.135
Squared multiple R: 0.018

Adjusted squared multiple R: 0.0
Standard error of estimate: 41.805

Effect	Coefficient	Std Error	t	P(2 Tail)
CONSTANT	51.952	21.341	2.434	0.018
CC1	0.431	0.885	0.488	0.628
CC2	0.432	0.491	0.880	0.382

Analysis of Variance

Source	SS	df	MS	F	P
Regression	2003.206	2	1001.603	0.573	0.567
Residual	108355.779	62	1747.674		

The comparison of the results from the two runs suggests that the approaches are equivalent. They result in the same $R^2 = 0.016$, the same standardized coefficients for the slope parameters, the same tolerance values, the same t values for the two slope parameters, and the same tail probabilities. Only the unstandardized coefficients, their standard errors, and the standard error of estimate are different, and so are the sum-of-squares values. The ANOVA results themselves have the same degrees of freedom, F values, and tail probabilities.

The next step involves inserting the ridge constant into the main diagonal of the correlation matrix. This can be performed using the following commands:

Command	Effect
Use Corrcc	Reads correlation matrix of variables CC1, CC2, and REC from file "Corrcc.sys"
Click Window, Worksheet	Correlation matrix is pulled into a window on the screen
Replace the first diagonal element, which currently is equal to 1, with 1.1; proceed until all diagonal elements are replaced	Substitutes correlations in diagonal of correlation matrix by 1 + ridge constants (see Table 11.2)
Click File, Save	Correlation matrix with altered entries in main diagonal is saved in file "CORRCC.SYS"

Using the altered correlation matrix we now can estimate the first set of ridge regression parameters. The ridge constant is $c = 0.1$. Parameters are estimated using the following commands:

Command	Effect
Use Corrcc	Reads correlation matrix of the variables CC1, CC2, and REC in which diagonal elements now contain ridge constant
Click Stats, MGLH, and Regression	Selects regression module from MGLH
Assign CC1 and CC2 to Independent and REC to Dependent	Specifies which variables are independent and which is dependent
Type "66" in rectangle for number of cases	Specifies sample size
Click Ok	Starts estimation of ridge regression parameters (notice again that there is no provision for an intercept); results of regression run appear on screen

The following output displays the results of this regression run:

```
MODEL REC = CC1+CC2 /N=65
>ESTIMATE
Dep Var: REC
N: 65
Multiple R: 0.123
Squared multiple R: 0.015

Adjusted squared multiple R: 0.0
Standard error of estimate: 1.049
```

Effect	Coefficient	Std Error	t	P
CC1	0.057	0.127	0.452	0.653
CC2	0.102	0.127	0.805	0.424

Analysis of Variance

Source	SS	df	MS	F	P
Regression	1.049	2	0.525	0.477	0.623
Residual	68.251	62	1.101		

This output gives the results of the estimation of ridge regression parameters for a ridge constant of $c = 0.1$. These results are equivalent to the ones in the first output, for which we had used Afifi and Clark's Dummy Observation Method. Specifically, the coefficients are the same. All other results differ because, in order to obtain a correct parameter estimate from the Dummy Observation Method, we had to artificially increase the sample size. As a consequence, the R^2 and all significance tests, including the ANOVA, cannot be interpreted in the first output. It still needs to be determined whether they can be interpreted here.

Using the correlation matrix method one can also create a series of estimates and chart a ridge trace. All that needs to be done is to replace the diagonal elements in the correlation matrix. Readers are invited to perform these replacements using ridge constants $0.1 \leq c \leq 1$ and to draw their own ridge trace for this example.

11.3.2 Least Median of Squares and Least Trimmed Squares Regression

To illustrate the use of LMS and LTS regression we use the software package S-Plus (Venables & Ripley, 1994). SYSTAT does not contain a module that would easily lead to solutions for these robust regression models.

S-Plus is an object-oriented system that provides a wide array of modules for robust estimation. Two ways of applying LMS and LTS regression using S-Plus will be illustrated. First, we show how to estimate and print regression parameters. Second, we show how to estimate regression parameters and to simultaneously draw the figure displayed in 11.6. To estimate regression parameters we issue the following commands. Each command is typed at the ">"-prompt, and is followed by striking the EN-TER key. It is important to note that S-Plus does distinguish between upper and lower case characters. Therefore, we use capitals only when needed.

Command	Effect
library(mass)	Invokes library MASS which contains the data file HILLS that we are using
attach(hills)	Makes variables in HILLS available by name
fm1 <- summary(lm(climb ~ time))	Estimates OLS regression of TIME on CLIMB; writes results to file FM1
fm2 <- lmsreg(time, climb)	Estimates LMS regression parameters; writes results to file FM2. Notice difference in variable order: The LM module expects the predictor first and then the criterion, separated by a tilde. The LMS and the LTS modules expect the criterion first and then the predictor, separated by a comma
	continued on next page

fm3 <– ltsreg(time, climb)	Estimates LTS regression parameters; writes results to file FM3
fm1	Sends contents of FM1 to screen; the screen is Windows' Notepad. Print commands have the same effect as "print screen" commands in DOS.
Click File, Print (repeat as needed)	Sends screen content to printer (page by page)
fm2	Sends contents of file FM2 to screen
Click File, Print (repeat as needed)	Sends screen contents to printer (page by page)
fm3	Sends contents of file FM3 to screen
Click File, Print (repeat as needed)	Sends screen contents to printer (page by page)

The following output displays the content of "FM1":

```
Call: lm(formula = climb ~ time)
Residuals:
   Min     1Q Median    3Q   Max
 -3227 -438.9 -76.41 549.6 1863

Coefficients:
               Value Std. Error  t value Pr(>|t|)
(Intercept) 307.3712 253.9620     1.2103   0.2348
       time  26.0549   3.3398     7.8012   0.0000

Residual standard error: 974.5 on 33 degrees of freedom
Multiple R-Squared: 0.6484
F-statistic: 60.86 on 1 and 33 degrees of freedom,
the p-value is 5.452e-009

Correlation of Coefficients:
```

```
     (Intercept)
time -0.7611
```

The following output displays parts of the content of "FM2":

```
$coef:
 Intercept              time
       650 -2.458929e-015

$resid:
  [1]     0  1850   250   150  2420
  [6]  2216  6850   150   150     0
 [11]  1450  1350  1550  -150   850
 [16]  2350  1550  -300   350   -50
 [21]  -350   850  1550   250   -50
 [26]  1350   150   300  1100  -150
 [31]  3750   -50  4550   200  4350

$wt:
  [1] 1 0 1 1 0 0 0 1 1 1 1 1 1 1 1 0 1 1
      1 1 1 1 1 1 1 1 1 1 1 1 0 1 0 1 0

$rsquared:
 [1] 0.51
```

The following output displays parts of the content of "FM3":

```
$coefficients:
 (Intercept)          x
    841.0253 -4.33677

$residuals:
  [1]  -121.28  1868.66   204.91   156.73  2499.01  2342.50
  [7]  7546.35   116.69    87.99   -18.64  2094.53  1345.67
 [13]  1640.86  -149.63   775.78  2472.31  1785.79  -149.94
 [19]   234.51   -99.79  -471.85   779.97  1565.55   136.75
 [25]  -160.00  1272.67   108.30   232.86  1127.98  -250.17
 [31]  3930.13  -100.59  5097.31   130.84  4852.13
```

```
$fitted.values:
 [1]  771.28  631.34  695.09  643.27  570.99  523.50  -46.35
 [8]  683.31  712.01  668.64    5.47  654.33  559.14  649.63
[15]  724.22  527.69  414.21  499.94  765.49  699.79  771.85
[22]  720.03  634.45  763.25  760.00  727.33  691.70  717.14
[29]  622.02  750.17  469.87  700.59  102.69  719.16  147.87
```

The first of these three outputs (FM1) presents the OLS results that we created using the LM module. The protocol shows first an overview of the residual distribution. It gives the smallest and the largest residual, and the three quartile points. In the center of the protocol there is the usual regression–ANOVA– type table. It shows parameter estimates, their standard errors, the t values (parameter estimates divided by their standard error), and the two-sided tail probabilities for the t values. This is followed by information on the residual standard error, the multiple R^2, and the F statistic for the ANOVA around the regression line. The last piece of information provided is the matrix of parameter intercorrelations. In the present example, this matrix contains no more than one correlation, for there are only two parameter estimates.

The second of these outputs (FM2) first presents information on the regression equation: It provides estimates of the intercept and the slope coefficients. What follows are the residuals for each of the races, presented in scale units of the dependent variable.

Most interesting is the string of weights that is printed below the residuals. A "1" indicates that a case can be considered relatively problem-free. In contrast, a "0" suggests that a case may be an outlier. The present example seems to contain six outliers.

In analogous fashion, the third output (FM3) first presents the parameter estimates. These are, as we indicated before, very close to the estimates provided by LMS regression. Residuals follow, also in units of the dependent scale. Because the parameters are so similar, the residuals are very similar also. The next block of information contains the fitted values.

The following example uses the same data and the same regression models. The goal is to show how to create a plot as displayed in Figure 11.6 using S-Plus. For the following commands to work, a Windows system and printer are required. Note again that S-Plus does distinguish between lower case and upper case letters.

Command	Effect
win.graph()	Invokes Windows graphics devices; presents a graph window
Click inside the command window	Carries you back to the command window
library(mass)	Makes files in library MASS available; these files contain the data file HILLS that we are using
attach(hills)	Makes variables in data file HILLS available by name
plot(climb, time, main = "OLS, LMS, and LTS Regression Lines for Hill Race Data")	Creates a scatterplot with variable Climb on the abscissa and variable Time on the ordinate; the string in quotation marks after "main" is the title of the graph
abline(lm(time $\sim$ climb), lwd = 1)	Creates a line whose parameters are provided by "lm", the OLS linear regression module; predictor is Climb, criterion is Time; lwd=1 specifies that the line be as wide as default
abline(lmsreg(climb, time), lwd = 1)	Same for LMS regression
abline(ltsreg(climb, time), lwd = 1)	Same for LTS regression
Click inside the graph window	Displays graph on screen
Click File, Print, Ok	Prints figure

Chapter 12

SYMMETRIC REGRESSION

This chapter presents various approaches to symmetric regression. This variant of regression has been discussed since Pearson (1901a) described a first symmetric regression model in a paper entitled "On Lines and Planes of Closest Fit to Systems of Points in Space." Symmetric regression models are beneficial for the following reasons:

1. The problems with inverse regression listed in Section 12.1.1 do not exist. Specifically, inverse regression, that is, regression of X on Y that starts from a point estimated from regressing Y on X, yields the original starting point. The inverse regression problem is solved because there is only one regression line.

2. Symmetric regression accounts for errors in the predictors. While there are many approaches to considering errors in predictors (the so-called errors-in-predictors models), none of these approaches also solve the inverse regression problem.

Thus, symmetric regression is useful because it solves both of these problems simultaneously. In addition, symmetric regression allows researchers to both estimate and test when there is no natural classification of variables in predictors and criteria.

This chapter first presents the Pearson (1901a) orthogonal regression solution. This is followed by two alternative solutions that have aroused interest particularly in biology and astrophysics, the bisector solution and the reduced major axis solution (see Isobe, Feigelson, Akritas, & Babu, 1990). A general model for OLS regression is presented next (von Eye & Rovine, 1995). The fourth section in this chapter introduces robust symmetrical regression (von Eye & Rovine, 1993). The last section discusses computational issues and presents computer applications.

12.1   Pearson's Orthogonal Regression

In many applications researchers are interested in describing the relationship between two variables, X and Y, by a regression line. In most of these applications there is a natural grouping of X and Y into predictor and criterion. For instance, when relating effort and outcome, effort is naturally the predictor and outcome is the criterion. However, in many other applications this grouping is either not plausible or researchers wish to be able to predict both Y from X and X from Y. Consider the relationship between weight and height that is used in many textbooks as an example for regression (e.g., Welkowitz, Ewen, & Cohen, 1990). It can be meaningful to predict weight from height; but it can also be meaningful to predict height from weight. Other examples include the prediction of the price for a car from the number of worker-hours needed to produce a car. This is part of a manufacturer's cost calculation. It can be equally meaningful to estimate worker-hours needed for production from the price of a car. This is part of the calculations performed by the competition.

The problems with inverse regression render these estimations problematic. Predictors measured with error make these problems even worse.

Therefore, the problem that is addressed in this chapter is different than the standard regression problem addressed in the first chapters of this book. Rather than estimating parameters for the functional relationship $E(Y|X)$, we attempt to specify an intrinsic functional relationship between X and Y. As we pointed out before, this is important particularly when the choice between X and Y as predictor is unclear, arbitrary, or ambiguous, and when both variables are measured with error.

Pearson (1901a) proposed to still use ordinary least squares methods

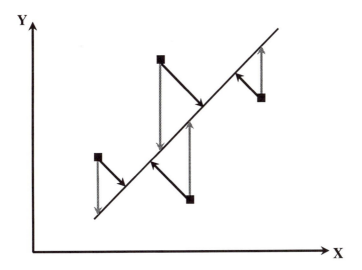

Figure 12.1: Illustration of residuals expressed in units of Y (gray) and perpendicular to the regression line (black).

for parameter estimation, but to employ an optimization criterion that is different than the standard sum of squared residuals. More specifically, Pearson proposed minimizing

$$\sum_i p_i^2, \tag{12.1}$$

where p_i is the ith case's perpendicular distance to the regression line, instead of the usual

$$\sum_i e_i^2 = \sum_i (y_i - \hat{y}_i)^2. \tag{12.2}$$

To illustrate the difference between (12.1) and (12.2) consider the four X–Y coordinate pairs (=data points) presented in Figure 12.1. The figure displays the four coordinate pairs, an imaginary regression line, and the residuals, defined using (12.1) in black, and defined using (12.2) in gray.

As is obvious from Figure 12.1, the black, perpendicular residuals are always shorter than the gray residuals parallel to the Y-axis. When the regression slope is $0 < b_1 < \infty$, there is no exception to this for residuals greater than zero. More important than this difference, however, is the difference in interpretation between the two definitions. In order to measure the magnitude of residuals, e_i, that are defined in units of Y, one only needs Y. In contrast, when measuring the magnitude of residuals p_i from orthogonal regression, one needs information about both X and Y.

More specifically, residuals in standard, asymmetric regression are defined by (12.2). Residuals in Pearson's orthogonal regression are defined as the perpendicular distance of point (x, y) to the regression line (see Pearson, 1901a, 1901b; von Eye & Rovine, 1995),

$$p_i = (y - \tan \theta (x - \bar{x}) - \bar{y}) \cos \theta,$$

where θ is the angle of the regression line with the Y-axis. The sum of the squared distances that is to be minimized is

$$\sum_i p_i^2 = \sum_i (y - \tan \theta (x - \bar{x}) - \bar{y})^2 \cos^2 \theta. \qquad (12.3)$$

Setting the first partial derivative with respect to θ equal to 0 yields the following solution for θ:

$$\tan 2\theta = \frac{2 \sum_i (y_i - \bar{y})(x_i - \bar{x})}{\sum_i (y_i - \bar{y})^2 - \sum_i (x_i - \bar{x})^2}. \qquad (12.4)$$

This is identical to the Pearson (1901a) solution (see (12.5)). Section 12.3 relates this solution to the standard, asymmetric OLS solution. As is obvious from intuition and Figure 12.1, the minimum of (12.3) gives the smallest total of all possible distance lines.

Another way of arriving at the solution given in (12.4) was proposed by Pearson (1901a). Pearson starts from considering an ellipse of the contours of the correlation surface of a two-dimensional data cloud. Using the centroid of the ellipse as the origin, the ellipse can be described as follows:

$$\frac{X^2}{\sigma_x^2} + \frac{Y^2}{\sigma_y^2} - \frac{2 r_{xy} xy}{\sigma_x \sigma_y} = 1.$$

Pearson showed that the ellipse of the contours of the correlation surface and the ellipse that describes the cloud of residuals have main axes that are orthogonal to each other. From this relation Pearson concluded that the best fitting line in the sense of (12.3) coincides in direction with the major axis of the ellipse of the correlation surface. The tangent of the angle, 2θ, of this axis is given by

$$\tan 2\theta = \frac{2r_{xy}\sigma_x\sigma_y}{\sigma_x^2 - \sigma_y^2}. \tag{12.5}$$

This is identical to (12.4). The mean squared of the residuals, p_i, is

$$\text{MSE}^2 = \frac{X^2Y^2}{\cot^2\theta}.$$

The following data example (von Eye & Rovine, 1995) relates psychometric intelligence and performance in school to each other. In a sample of $n = 7$ children, the following IQ scores were observed: IQ = $(90, 92, 93, 95, 97, 98, 100)$. The performance scores are, in the same order, P = $(39, 42, 36, 45, 39, 45, 42)$. The Pearson correlation between these two variables is $r = 0.421$. Regressing *Performance* on IQ yields the following OLS solution using (12.2) as the criterion to be minimized:

Performance = $3.642 + 0.395 * \text{IQ} + \text{Residual}$.

Substituting predictor for criterion and criterion for predictor yields

IQ = $76.538 + 0.449 * \text{Performance} + \text{Residual}$.

The data points and these two regression lines appear in Figure 12.2, where the regression for *Performance* on IQ is depicted by the thicker line, and the regression of IQ on *Performance* is depicted by the thinner line.

The following paragraphs illustrate problems with inverse regression that are typical of standard, asymmetric regression where one estimates parameters for two regression lines.

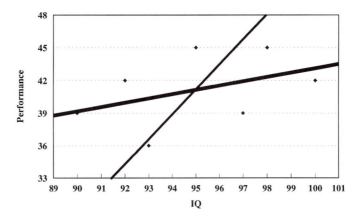

Figure 12.2: Regression lines for Performance on IQ (thicker line) and IQ on Performance (thinner line).

12.1.1 Symmetry and Inverse Prediction

One of the most striking and counterintuitive results from employing two asymmetric regression lines for prediction and inverse prediction is that back-prediction does not carry one back to the point where the predictions originated. This result is illustrated using the data in Figure 12.2.

Consider the following sequence of predictions and inverse predictions:

$$(1) \quad x_i \to \hat{y}_i, \quad \text{and}$$
$$(2) \quad \hat{y}_i \to x_i.$$

This sequence describes *symmetric* prediction if

$$\hat{x}_i = x_i. \tag{12.6}$$

This applies accordingly if predictions start from some value y_i. If, however,

$$\hat{x}_i \neq x_i. \tag{12.7}$$

this sequence describes *asymmetric* prediction. Table 12.1 presents point predictions and inverse point predictions from the above regression equa-

Table 12.1: *Point Predictions and Inverse Point Predictions Using Asymmetric OLS Regression*

Predictor Value	Estimate	Back Estimate	
IQ	Performance	IQ	Difference
70	31.29	90.59	20.59
80	35.24	92.36	12.36
90	39.12	94.14	4.14
100	43.12	95.90	−4.1
110	47.09	97.68	−12.32
120	51.04	99.46	−20.54
130	54.99	101.23	−28.77
Performance	IQ	Performance	Difference
24	87.31	38.13	14.13
30	90.01	39.20	9.2
36	92.70	40.26	4.26
42	95.39	41.32	−0.68
48	98.09	42.39	−5.61
54	100.78	43.45	−10.55
60	103.48	44.52	−15.48

tions.

The top panel in Table 12.1 displays estimates of *Performance* (second column) as predicted from *IQ* (first column) using the above regression equation. The third column of this panel contains the back estimates of *IQ*. These estimates were calculated using the regression equation (12.6), with the estimates of *Performance* as starting values. The fourth column displays the differences between back estimates and starting values. The bottom panel of Table 12.1 displays results for the prediction of *IQ* from *Performance* (Columns 1 and 2). Column 3 of the bottom panel displays the back-estimated *Performance* scores, where the *IQ* estimates served as predictor scores. The last column contains the differences between back-estimated and original *Performance* scores.

The results in Table 12.1 and other considerations suggest that

1. The differences between starting values and back-estimated starting values increase with the distance from the predictor mean;

2. Differences increase as the correlation between predictor and criterion decreases; and

3. Differences increase as regression slopes become flatter.

When researchers only wish to predict Y values, this asymmetry may not matter. However, there are many applications in which this characteristic can become a problem. Consider the following example. A teacher has established the number of learning trials needed for a student to reach a criterion. A new student moves to town. This student performs below criterion. Using inverse regression the teacher estimates the number of make-up sessions needed for this student to reach the criterion. Depending on the strength of the relationship, the inverse prediction may not only be far away from the number of sessions the new student may actually need to reach the criterion. The estimate of lessons needed may indicate a number of sessions that is impossible to implement and, thus, makes it very hard for the student to be considered an adequate performer.

On another note, it should be noticed that the above example includes discrepancies that are so large that, in other contexts, they would qualify as statistically significant.

12.1.2   The Orthogonal Regression Solution

Inserting into (12.4) creates an estimate for the slope parameter of Pearson's orthogonal regression. More specifically, we estimate the tangent of two times the angle θ of the orthogonal regression line as

$$\tan 2\theta = \frac{2 * 0.421 * 3.559 * 3.338}{3.559^2 - 3.338^2} = 6.54.$$

From this value we calculate $2 * \theta = 81.31°$ and $\theta = 40.65°$. Thus, the angle of the symmetrical regression line is $40.65°$. As in standard, asymmetric OLS regression, the symmetric regression line goes through the centroid (center of gravity, the mean) of the data cloud. The centroid has for coordinates the arithmetic means of predictor and criterion. In the present example, we obtain for the centroid

$$C = \left(\begin{array}{c} 95.00 \\ 41.13 \end{array} \right).$$

Straight lines can be sufficiently described by a point and the angle of the slope. Thus, we have a complete solution. This solution has the following characteristics:

1. It is an OLS solution in that it minimizes the sum of the squared perpendicular distances of data points from the regression line.

2. It is symmetric in the sense specified in Equations (12.6) and (12.7).

3. As a result of 2, a prediction of Y from X, followed by an inverse prediction that originates from the predicted y value, carries one back to the original X value. This applies accordingly when the prediction starts from Y.

4. This method is applicable in particular when both predictor and criterion are measured with error.

5. This method is applicable in particular when there is no natural grouping of variables into predictors and criteria.

6. The Pearson orthogonal symmetrical regression and the OLS asymmetric regression lines coincide when $|r| = 1.0$.

7. The Pearson orthogonal symmetrical regression line is identical to the first principal component in principal component analysis (see, e.g., Pearson, 1901a, 1901b; Morrison, 1990). The second principal component has the same orientation as the main axis of the ellipse for the residuals, p_i, in orthogonal regression, and so forth.

Figure 12.3 displays the scatterplot of the *IQ-Performance* data, the two asymmetric regression lines, and the symmetric regression line (thick gray). Note that the symmetric regression line does not half the angle between the two asymmetric regression lines.

12.1.3 Inferences in Orthogonal Regression

Based on the assumption that predictor and criterion are jointly normally distributed, Jolicoeur (1973) and Jolicoeur and Mosiman (1968) proposed an approximate method for estimating a confidence interval around the

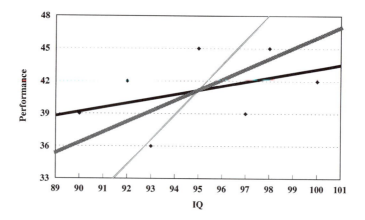

Figure 12.3: Three regression lines for Performance – IQ data.

slope coefficient of orthogonal regression. Let κ be the slope parameter and

$$k_1 \leq \kappa \leq k_2$$

the confidence interval. Then, the limits, k_1 and k_2, are

$$k_1 = \frac{s_y^2 - s_x^2 + \sqrt{(s_y^2 - s_x^2)^2 + 4s_{xy}^2 - 4Q}}{2(s_{xy} + \sqrt{Q})}$$

and

$$k_1 = \frac{s_y^2 - s_x^2 - \sqrt{(s_y^2 - s_x^2)^2 + 4s_{xy}^2 - 4Q}}{2(s_{xy} + \sqrt{Q})},$$

where

$$Q = \frac{F_{1-\alpha;1;n-2}(s_x^2 s_y^2 - s_{xy}^2)}{n - 2},$$

where s_x and s_y are the variable standard deviations and s_{xy} is the co-variance of X and Y.

This approximation is reasonably close to the nominal α even if the correlation is as low as $\rho = 0.4$ and the sample size is as small as $n = 10$. Notice, however, that the values of k_1 and k_2 can be imaginary if the expression under the square root in the numerator becomes negative. This is most likely when the correlation between X and Y is small. Notice in addition that, as is usual with confidence intervals, the values of k_1 and k_2 are always real and assume the same sign as the correlation if the correlation is statistically significant.

12.1.4 Problems with Orthogonal Regression

The following two problems are apparent for orthogonal regression:

1. Lack of invariance against specific types of linear transformation. When X and Y are subjected to linear transformations with different parameters, the parameter estimates for orthogonal regression change. This is the case if, for example, X is replaced by $b_x X$ and Y is replaced by $b_y Y$, with $b_x \neq b_y$. This is not the case in standard, asymmetric regression.

2. Dependence of estimates on variance and mean. Parameter estimates of orthogonal regression tend to be determined to a large extent by the variable with the larger variance and the greater mean.

   Logarithmic transformation and standardization have been proposed as solutions for both problems (Jolicoeur, 1991) (see the reduced major axis solution in the following section).

12.2 Other Solutions

Fleury (1991) distinguishes between two forms of regression: Model I and Model II. Model I Regression of Y on X is asymmetric and minimizes (12.2), that is, the sum of the squared vertical differences between data points and the regression line. If X is regressed on Y, Model I Regression minimizes the sum of the squared horizontal differences between data points and the regression line (see Figure 12.4, below). Model II Regression is what we term symmetric regression. It has a number of variants, three of which are presented in this volume (for additional least squares

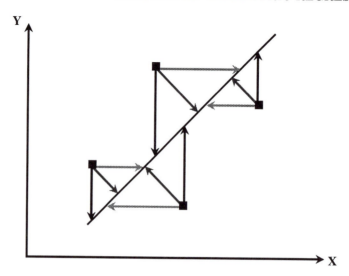

Figure 12.4: Illustration of goal functions of orthogonal regression and reduced major axis regression.

solutions see Isobe et al., 1990). The first is Pearson's orthogonal regression, also called major axis regression (see Section 12.1). The second is reduced major axis regression (Kermak & Haldane, 1950), also called *impartial regression*, (see Strömberg, 1940). The third is *bisector regression* (Rubin, Burstein, & Thonnerd, 1980), also called *double regression*, see Pierce and Tulley (1988). The second and the third solutions are reviewed in this section.

As was illustrated in Figure 12.1, Pearson's major axis solution minimizes the sum of the squared perpendicular distances between the data points and the regression line. In contrast, the reduced major axis regression minimizes the sum of the areas of the right triangles created by the data points and the regression line. The two approaches of major and reduced major regression are illustrated in Figure 12.4. Standard, asymmetric regressions of Y on X and X on Y are also illustrated in the figure.

The black arrows in Figure 12.4 illustrate the standard residual definition for the regression of Y on X. The estimated y value (regression line) is subtracted from the observed y value (squares) (see (12.2)). The sum of these squared differences is minimized. In an analogous fashion, the light gray arrows illustrate the regression of X on Y, where residuals are expressed in units of X.

Pearson's orthogonal regression minimizes the sum of the squared dark gray differences in Figure 12.4. The dark gray arrows are perpendicular to the regression line. Thus, they represent the shortest possible connection between an observed data point and the regression line.

The two asymmetric regression lines and the symmetric orthogonal regression line share in common that they minimize lines that connect data points and the regression line. The reduced major axis solution reduces an area. Specifically, it reduces the area of the right triangle spanned by the black, vertical arrow; the light gray, horizontal arrow; and the regression line. This triangle is a right triangle because the angle between the vertical and the horizontal arrow is a right angle.

The equation for the reduced major axis regression is

$$\hat{y}_i = \bar{y} + sgn(r)\frac{s_y(x_i - \bar{x})}{s_x},$$

where $sgn(r)$ is the sign of the correlation between X and Y. (A definition of the area of the triangle is given in Section 12.5.1 on computational issues.) The slope parameter for the reduced major axis solution is

$$b_1 = sgn(s_{xy})\sqrt{\hat{\beta}_1\hat{\beta}_2}, \tag{12.8}$$

where $\hat{\beta}_1$ and $\hat{\beta}_2$ are the estimates of the coefficients for the regression of Y on X and X on Y, respectively.

The approximate limits of the confidence interval for the slope coefficient in (12.8) are

$$k_1 = sgn(r)\frac{s_y(\sqrt{B+1} - \sqrt{B})}{s_x}$$

and

$$k_2 = sgn(r)\frac{s_y(\sqrt{B+1} + \sqrt{B})}{s_x},$$

where

$$B = \frac{F_{1-\alpha;1;n-2}(1 - r^2)}{n - 2}.$$

The main advantage of the reduced major axis solution over the major axis solution is that it is scale-independent. Nevertheless, there are recommendations to use this method only if

1. the sample size is $n \geq 20$

2. the correlation between X and Y is assumed to be strong $\rho \geq 0.6$

3. the joint distribution of X and Y is bivariate normal (Jolicoeur, 1991).

Isobe et al. (1990) urge similar cautions. These cautions, however, are not proposed specifically for the reduced major axis solution, but in general, for all regression solutions, including all asymmetric and symmetric solutions included in this volume.

The last method for symmetric regression to be mentioned here is called the *bisector solution*. The bisector solution proceeds in two steps. First, it estimates the parameters for the two asymmetric regressions of Y on X and X on Y. Second, it halves the angle between these two regression lines; that is, it halves the area between the regression lines. The slope coefficient for the bisector solution is

$$b_1 = \frac{\hat{\beta}_1\hat{\beta}_2 - 1 + \sqrt{(1 + \hat{\beta}_1^2)(1 + \hat{\beta}_2^2)}}{\hat{\beta}_1 + \hat{\beta}_2},$$

where, as in (12.8), the $\hat{\beta}$ are the estimates of the standard OLS asymmetric regression slope parameters.

12.2.1 Choice of Method for Symmetric Regression

Isobe et al. (1990) performed simulation runs to investigate the behavior of the two asymmetric regression solutions, the bisector solution, the reduced major axis solution, and the major axis solution. They conclude that the bisector solution can be recommended because its standard deviations are the smallest. The Pearson orthogonal regression solution displays, on average, the largest discrepancies between theoretical and estimated values.

When applying regression, one goal is to create an optimal solution. Minimal residuals and minimal standard deviations are examples of criteria for optimal solutions. In addition, significance testing and confidence intervals are of importance. There are solutions for significance testing for the Pearson orthogonal regression and the reduced major axis solution. These solutions were described earlier in this chapter. Therefore, the following recommendations can be given as to the selection of methods for symmetric regression:

- When estimation of scores is of interest rather than statistical significance of parameters, the bisector solution may be the best, because it creates solutions with the smallest standard deviations. Examples of applications in which predominantly estimation is important include the estimation of time, amount of work needed to reach a criterion, and distance. All this applies in particular when there is no obvious classification of variables in predictors and criteria, and when both predictors and criteria are measured with error.

- When estimation of confidence intervals and significance testing are important, the pragmatic selection is either Pearson's orthogonal regression or the reduced major axis solution. Either solution can be calculated using standard statistical software. Section 12.5 illustrates this using the statistical software package SYSTAT. These solutions may not produce the smallest standard deviations. However, they do not require time-consuming bootstrapping or other procedures that yield estimates for standard errors of parameter estimates. Thus, whenever researchers have available software packages that allow them to perform steps of the type illustrated in Section 12.5, they can create a complete solution that (1) differs from the solution with the smallest standard deviation only minimally, (2) allows for significance testing and estimation, and (3) carries all the benefits of symmetric regression.

The following data example uses data from Finkelstein et al. (1994). A sample of $n = 70$ adolescent boys and girls answered a questionnaire

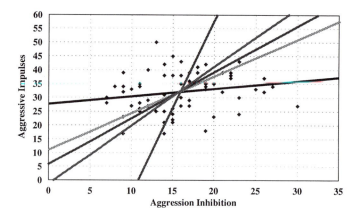

Figure 12.5: Five regression solutions for aggression data.

that included, among others, the scales Aggressive Impulses, AI, and Aggression Inhibition, AIR. An Aggressive Impulse is defined as the urge to commit an aggressive act against another person. Aggression Inhibition is defined as a mental block that prevents one from committing aggressive acts. We analyze these two scales using the five regression models

1. AIR on AI

2. AI on AIR

3. major axis (orthogonal)

4. reduced major axis

5. bisector

Figure 12.5 displays the data and the five regression lines.

The correlation[1] between Aggressive Impulses and Aggression Inhibition is $r = 0.205$. The figure suggests accordingly that the angle between the regression line of AIR on AI and the regression line of AI on AIR

---

[1]This correlation is statistically not significant ($t = 1.727; p = 0.089$). Thus, application of symmetric regression is counter the recommendations given by several authors. However, one of the main purposes of the present example is to dramatize the differences between asymmetric and symmetric regression. There is no intention to make a substantive statement concerning the relationship between AI and AIR.

is wide. These are the most extreme regression lines in the figure, that is, the lines with the steepest and the flattest slopes. As a result, inverse regression that starts from a predicted value can carry us far away from where the predictions originated.

All symmetric regression solutions provide regression lines between the two asymmetric lines. The differences between the symmetric lines are small. Each of the differences is smaller than any of the standard errors (for more technical detail see Isobe et al. (1990); for an application to astronomy data see Feigelson and Babu (1992) and von Eye and Rovine (1993)). The slopes of the five solutions are as follows:

1. 0.276
2. 6.579
3. 0.307
4. 0.742
5. 0.889

All of these regression lines go through the center of gravity (centroid) of the bivariate distribution. Thus, the centroid and the slopes give a complete description of the regression lines.

12.3   A General Model for OLS Regression

This section presents a general model for OLS regression (Jolicoeur, 1991; von Eye & Rovine, 1995). This model unifies the approaches presented in the previous sections and in Figure 12.4. The two asymmetric approaches and the major axis approach will be identified as special cases of this unified model. The nature of this section differs from the others in that it does not present a method and its applications. Rather, it presents and explains this unified model. The benefit from reading this section is, therefore, that readers gain a deeper understanding of the methods of symmetric regression. There is no immediate benefit in the sense that calculations become easier, faster, or possible.

The General Model is presented using means of planimetry, that is, two-dimensional geometry. Figure 12.6 gives a representation of the components of the model.

Figure 12.6 provides a more general representation of the problem

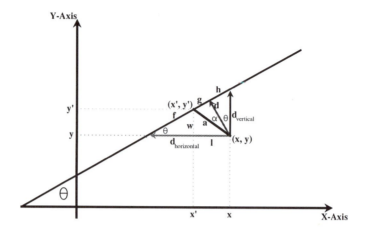

Figure 12.6: Illustration of a general model for OLS regression.

already represented in Figure 12.4. The three solutions depicted in the figure share in common that they determine parameters of a regression line by minimizing the distances, d, of the residuals from this regression line. Figures 12.4 and 12.6 show three examples of such distances. There is first the light gray, horizontal arrows that depict distances defined in units of the X-axis. This distance or residual definition is used when regressing X on Y. Second, there are black, vertical arrows. This distance or residual definition is used when regressing Y on X. Both of these approaches are asymmetric. Third, there are the dark gray arrows that are perpendicular to the regression line. This approach is used in Pearson's orthogonal regression solution.

From the unified model perspective, one can state a more general regression problem as follows (von Eye & Rovine, 1995): what is the regression line obtained by minimizing the sum of the squared distances between observed data points (x, y) and an arbitrary data point (x', y') at some angle α from the perpendicular intersection of the regression line? In Figure 12.6 a regression line is drawn that forms an angle, θ, with the X-axis. This regression line goes through the centroid of the bivariate distribution of X and Y. The angle α represents the deviation from the perpendicular of the distance to be minimized. The figure also shows data point (x, y) and an arbitrary data point (x', y'). It shows the angles

α and θ and three sample definitions of the distance between (x, y) and the regression line: d for the perpendicular distance (dark gray arrow), $d_{horizontal}$ for the horizontal distance (light gray arrow), and $d_{vertical}$ for the vertical distance (black arrow). Distance a is an arbitrary distance. It lies within the triangle minimized by the reduced major axis solution. However, it does not coincide with any of the distances minimized for the two asymmetric and the major axis solutions.

Now, to answer the above question consider, as in Section 12.1.2, the regression line that, at angle θ, goes through the centroid,

$$y_i - \bar{y} = \tan \theta (x_i - \bar{x}).$$

The distance to be minimized is denoted by the black solid line labeled a in Figure 12.6. When minimizing this distance consider the triangle formed by the three lines a, w, and l, where l is a part of $d_{horizontal}$ with $l \le d_{horizontal}$. The distance a can be expressed as

$$a = \sqrt{w^2 + l^2}.$$

The vertical distance from the point (x, y) to the regression line is

$$d_{i,vertical} = (y - \tan \theta (x_i - \bar{x})) - \bar{y}.$$

The perpendicular distance (dark gray line) of (x, y) to the regression line is

$$d_i = (y - \tan \theta (x_i - \bar{x}) - \bar{y}) \cos \theta.$$

Setting[2]

$$a = d \sec \alpha,$$

and

$$
\begin{aligned}
l &= a \cos(90 - \alpha - \theta) \\
w &= a \sin(90 - \alpha - \theta)
\end{aligned}
$$

---

[2]In the equation "sec" is short for secant. In a right triangle a secant can be defined as the ratio of the hypotenuse to the side adjacent to a given angle.

we can describe the coordinates of point (x', y') as

$$
\begin{aligned}
x' &= x - l \\
y' &= y + w.
\end{aligned}
$$

Using these terms, we can give a more explicit specification of the distance to be minimized. It is

$$
a_i = (y - \tan\theta(x_i - \bar{x} - \bar{y})^2(\cos^2\theta\sec^2\alpha)) \\
* (\cos^2(90 - \alpha - \theta) + sin^2(90 - \alpha - \theta)).
$$

OLS minimizes the sum of the squared distances, that is,

$$
\sum_i (y_i - \tan\theta(x_i - \bar{x}) - \bar{y})^2 \cos^2\theta\sec^2\alpha. \tag{12.9}
$$

In the following paragraphs we show that the two asymmetric regression solutions and Pearson's orthogonal solution are special cases of Equation (12.9). First, we consider the regression of Y onto X. For this case, $\alpha = \theta$ and the function to be minimized is

$$
\sum_i (y_i - \tan\theta(x_i - \bar{x}) - \bar{y})^2.
$$

Taking the first partial derivative with respect to θ and setting that to 0 yields the solution for θ in the form of

$$
\tan\theta = \frac{\sum_i(y_i - \bar{y})(x_i - \bar{x})}{\sum_i(x_i - \bar{x})^2}.
$$

This is identical to the solution for the regression of Y onto X.

Second, we consider the orthogonal, symmetrical solution, that is, the major axis solution. Here, $\alpha = 0$, and the equation to be minimized is

$$
\sum_i (y_i - \tan\theta(x_i - \bar{x}) - \bar{y})^2 \cos^2\theta.
$$

The first partial derivative of this expression with respect to θ, set to 0, is

$$\tan 2\theta = \frac{2\sum_i (y_i - \bar{y})(x_i - \bar{x})}{\sum_i (y_i - \bar{y})^2 - \sum_i (x_i - \bar{x})^2}.$$

This is Pearson's solution.

Third, we consider the regression of X onto Y. For this asymmetric regression $\alpha = 90° - \theta$, and the equation to be minimized is

$$\sum_i (y_i - \tan\theta(x_i - \bar{x}) - \bar{y})^2 \cot^2\theta.$$

The first partial derivative of this expression with respect to θ, set to 0, is

$$\cot\theta = \frac{\sum_i (y_i - \bar{y})(x_i - \bar{x})}{\sum_i (y_i - \bar{y})^2}.$$

The derivations for these three results are given in von Eye and Rovine (1995).

12.3.1 Discussion

From a data analyst's perspective, the three solutions presented here can be viewed as different hypotheses about the nature of residuals. The solutions discussed here cover all regression lines between $\alpha = \theta$ and $\alpha = 90 - \theta$. As the regression line sweeps from one extreme to the other, the degree to which X and Y contribute to the slope of the line changes. l moves from $d_{horizontal}$ to 0, and w moves from 0 to $d_{vertical}$ (see Figure 12.6).

The concept of multiple symmetrical regression, also introduced by Pearson (1901a), is analogous to the concept of simple symmetrical regression. Consider the three variables A, B, and C. The data cloud, that is, the joint distribution of these variables, can be described by a three-dimensional ellipsoid. This ellipsoid has three axes. As in simple, that is, bivariate regression, the solution for three-dimensional symmetrical regression is a single regression line. In three-dimensional asymmetric regression the solution is no longer a line, it is a (hyper)plane. The best fitting symmetrical regression line goes through the centroid and has the same orientation, that is, slope, as the main axis of the ellipsoid. This applies accordingly to more than three dimensions (see Rollett, 1996).

12.4 Robust Symmetrical Regression

Outliers tend to bias parameter estimates. This applies in particular to estimates from OLS. The above symmetric regression methods use OLS methods. Therefore, they are as sensitive to outliers as standard asymmetric regression methods that also use OLS methods. To create a regression solution that is both symmetric and robust, von Eye and Rovine (1993, 1995) proposed combining Pearson's orthogonal regression method with Rousseeouw's LMS regression. This method proceeds as follows

1. Select a subsample of size $n_j \leq n$ and calculate the slope of the orthogonal regression line and the median of the squared residuals where residuals are defined as in Pearson's solution; save the parameter estimates and median.

2. Repeat Step 1 until the smallest median has been found.

The method thus described, robust symmetric orthogonal regression, shares all the virtues of both orthogonal and LMS regression. It also shares the shortcomings of both (see Section 12.1.4).

12.5 Computational Issues

In spite of its obvious benefits and its recently increased use in biology and astrophysics, symmetric regression is not part of standard statistical software packages. However, for some of the solutions discussed in this chapter there either are special programs available or solutions can be easily implemented using commercial software. For instance, Isobe and Feigelson (Isobe et al., 1990) make available copies of a FORTRAN 77 program that calculates the five regression models discussed in their 1990 paper. The present section illustrates the calculation of the major axis and the reduced major axis solutions using SYSTAT (Fleury, 1991).

As in Section 12.5 we assume the reader is familiar with standard data and file manipulation routines in SYSTAT. Therefore, we focus on the methods of symmetric regression, using a data file already in existence as a SYSTAT system file. Specifically, we use file AGGR.SYS, which contains the data for the aggression data example (see Figure 12.5). First, we show how to create a major axis solution. Then, we show how to create a reduced major axis solution (Section 12.2.1). Both SYSTAT solutions are based on Fleury (1991).

12.5.1   Computing a Major Axis Regression

The NONLIN module in SYSTAT allows one to estimate parameters in nonlinear estimation problems. It is important for the present context that this module allows one to specify both the functions to be minimized (loss function) and the minimization criterion. The default criterion is least squares. Therefore, we only need to specify the loss function. For the major axis solution the loss function is

$$\text{LOSS} = \frac{(y - (b_0 + b_1 x))^2}{1 + b_1^2}.$$

Parameter estimates for b_0 and b_1 can be obtained as follows.

Command	Effect
Use Aggr (AI, AIR)	Reads variables AIR and AI from file "Aggr.sys;" SYSTAT presents list of variables on screen
Click Stats, Nonlin	Invokes NONLIN module in SYSTAT
Click Loss Function	Opens the window in which one can specify Loss Function
Type into the field for the Loss function the expression "(AI87 - (B0 +B1* AIR87))^2/(1 + B1^2)"	Specifies loss function for major axis solution; parameters to be estimated are b0 (intercept) and b1 (slope)
	continued on next page

Click OK, Stats, NONLIN, Model	Invokes the NONLIN module again; opens the window in which we specify the regression model
Type in the field for the regression model "AI87 = B0 + B1*AIR87"	Specifies the regression model and its parameters
Click OK	Starts iterations; SYSTAT presents overview of iteration process on screen, followed by the parameter estimates; carries us back to the SYSTAT command mode window

After these operations we have the iteration protocol and the result of the iterations on screen. Highlighting all this and invoking the Print command of the pull-down File menu gives us the following printout:

```
LOSS  = (ai87-(b0+b1*air87))^2/(1+b1^2)
```

```
>ESTIMATE
```

```
 Iteration
No.      Loss       B0            B1
   0 .177607D+04 .111633D+02 .152146D+00
   1 .173618D+04 .642059D+01 .300026D+00
   2 .173611D+04 .620045D+01 .306890D+00
   3 .173611D+04 .619920D+01 .306929D+00
   4 .173611D+04 .619920D+01 .306929D+00
```

```
Dependent variable is AI87
```

```
Final value of loss function is     1736.106
Zero weights, missing data or estimates reduced
degrees of freedom
                                    Wald Conf. Interval
Parameter  Estimate   A.S.E.  Param/ASE  Lower < 95%> Upper
BO          6.199     4.152    1.493     -2.085       14.484
B1          0.307     0.128    2.399      0.052        0.562
```

The first line in this output contains the specification of the loss function. The second line shows the command "estimate." What follows is an overview of results from the iteration process. The program lists the number of iterations, the value assumed by the loss function, and the parameter values calculated at each iteration step. The values given for Iteration 0 are starting values for the iteration process. It should be noticed that numerical accuracy of the program goes beyond the six decimals printed. This can be concluded from the last three iteration steps. Although the first six decimals of the loss function do not change, the parameter values do change from the third to the fourth iteration step.

After this protocol the program names the dependent variable (which, in some applications of symmetrical regression, may be a misnomer) and the final value of the loss function. This value is comparable to the residual sum of squares. The following line indicates that cases had been eliminated due to missing data. The final element of the output is the listing of the parameter estimates. The slope parameter estimate is the same as the one listed at the end of Section 12.2.1.

12.5.2 Computing Reduced Major Axis Solution

The major axis and the reduced major axis OLS solutions differ only in their loss function. The loss function for the reduced major axis solution is

$$\text{LOSS} = \frac{(y - (b_0 * b_1 x))^2}{|b_1|}.$$

The following commands yield parameter estimates for the reduced major axis solution.

Command	Effect
Use Aggr (AI, AIR)	Reads variables AIR and AI from file "Aggr.sys". SYSTAT presents list of variables on screen
Click Stats, Nonlin	Invokes NONLIN module in SYSTAT
	continued on next page

Click Loss Function	Opens the window in which one can specify Loss Function
Type into the field for the Loss function the expression "(AI87 - (B0 + B1* AIR87))^2/abs(B1)"	Specifies Loss Function for Reduced Major Axis Solution; parameters to be estimated are b0 (intercept) and b1 (slope)
Click OK, Stats, NONLIN, Model	Invokes the NONLIN module again; opens the window in which we specify the regression model
Type in the field for the regression model "AI87 = B0 + B1*AIR87"	Specifies the regression model and its parameters
Click OK	Starts iterations; SYSTAT presents overview of iteration process on screen, followed by the parameter estimates; carries us back to the SYSTAT command mode window

The following output presents the protocol created from these commands:

```
LOSS  = (ai87-(b0+b1*air87))^2/abs(b1)

>ESTIMATE

 Iteration
 No.      Loss        B0           B1
    0 .618924D+04 .619920D+01 .306929D+00
    1 .481618D+04 .211888D+01 .434155D+00
    2 .422637D+04-.246146D+01 .576972D+00
    3 .407632D+04-.612368D+01 .691162D+00
    4 .406342D+04-.759789D+01 .737128D+00
    5 .406330D+04-.776169D+01 .742235D+00
    6 .406330D+04-.776341D+01 .742289D+00
    7 .406330D+04-.776341D+01 .742289D+00

Dependent variable is AI87
```

```
Final value of loss function is      4063.300
Zero weights, missing data or estimates reduced
degrees of freedom
```

```
                                      Wald Conf. Interval
Parameter  Estimate   A.S.E.   Param/ASE   Lower < 95%> Upper
BO          -7.763    3.726     -2.083    -15.199      -0.328
B1           0.742    0.114      6.540      0.516        0.969
```

This protocol has the same form as the one for the major axis solution. One obvious difference is that the iteration took longer to find the minimum of the loss function than that for the major axis solution. Parameter estimates are $b_0 = -7.763$ and $b_1 = 0.742$. These values are identical to the ones reported at the end of Section 12.2.1. It should be noticed that whereas the value assumed by the loss function is not invariant against linear transformations, the parameter estimates are. This can be illustrated by dividing the above loss function by 2 (Fleury, 1991). This transformation yields a final value of the loss function that is half that in the above output. The parameter estimates, however, remain unchanged.

To enable readers to compare the symmetric regression solutions with the asymmetric ones we include the protocols from the two asymmetric regression runs in the following output:

```
>MODEL AI87 = CONSTANT+AIR87
>ESTIMATE
44 case(s) deleted due to missing data.

Dep Var: AI87
N: 70
Multiple R: 0.205
Squared multiple R: 0.042

Adjusted squared multiple R: 0.028
Standard error of estimate: 5.169

Effect    Coefficient   Std Error    t     P(2 Tail)

CONSTANT     11.163        2.892    3.860    0.000
```

AIR87   0.152   0.088   1.727   0.089

Dep Var: AIR87
N: 70
Multiple R: 0.205
Squared multiple R: 0.042

Adjusted squared multiple R: 0.028
Standard error of estimate: 6.964

Effect   Coefficient   Std Error   t   P(2 Tail)

CONSTANT   27.641   2.697   10.249   0.000
AI87   0.276   0.160   1.727   0.089

Chapter 13

VARIABLE SELECTION TECHNIQUES

In this chapter we deal with the question of how to select from a pool of independent variables a subset which explains or predicts the dependent variable well enough so that the contribution of the variables not selected can be neglected or perhaps considered pure error. This topic is also known as "subset selection techniques". The two aims, explanation and prediction, are distinct in that an obtained regression equation which gives a good prediction might, from a theoretical viewpoint be not very plausible. As the techniques used for variable selection and prediction are the same, we will not further consider this distinction. The terms predictor and explanatory variable are therefore used interchangeably. However, the following remark should be kept in mind: If prediction is the focus, one can base variable selection predominantly on statistical arguments. In contrast, if explanation is the focus, theoretical arguments guide the variable selection process. The reason for this is that the so–called *F-to-enter* statistic, testing whether a particular regression coefficient is zero, does not have an F distribution if the entered variable is selected according to some optimality criterion (Draper, Guttman, & Kanemasu, 1971; Pope & Webster, 1972).

That variable selection poses problems is obvious if the regression model is used for explanatory purposes, as predictors are virtually always intercorrelated. Therefore, values of parameter estimates change when including or eliminating predictors. As a consequence, the interpretation of the regression model can change. Yet, there is another argument for not fitting a regression model with more variables than are actually needed. To see this, we first have to establish notation. Let

$$E(y) = \beta_0 + \beta_1 x_1 + \cdots + \beta_a x_a + \cdots + \beta_{a+b} x_{a+b}$$

be the regression equation for a single observation where the set of $a + b$ predictors is divided into two non-overlapping subsets, A and B, which contain the indices from 1 to a and from $a + 1$ to $a + b$, respectively. The OLS estimate for the β vector is given in matrix notation by

$$\hat{\beta} = (\mathbf{X'X})^{-1}\mathbf{X'y},$$

where $\mathbf{X}$ is the design matrix of all the $A + B$ predictors including a vector for the intercept, and $\mathbf{y}$ now denotes the vector of observations of the dependent variable. The reason for subdividing the predictors into two sets is that we can now write the design matrix as $\mathbf{X} = (\mathbf{X_A}, \mathbf{X_B})$ and can express the OLS estimate of the regression coefficients after selecting a subset of predictors for the regression as

$$\hat{\beta}_A = (\mathbf{X'_A X_A})^{-1}\mathbf{X'_A y},$$

tacitly assuming, of course, that the selected variables are reordered so that they are contained in A.

With this notation the argument for not fitting a regression model with too many variables is validated sin it can be shown that the following inequality holds (for a proof see Miller, 1990):

$$\mathrm{var}(\boldsymbol{x'}\hat{\beta}) \geq \mathrm{var}(\boldsymbol{x'_A}\hat{\beta}_A).$$

In words, this inequality states, that if we base our prediction on a subset of all available predictors the variability of the predicted value, $\hat{\mathbf{y}}_\mathbf{A} = \boldsymbol{x'_A}\hat{\beta}_A$, is generally reduced compared to the variability of a prediction from the complete set, $\hat{\mathbf{y}} = \boldsymbol{x'}\hat{\beta}$. Hence, the precision of our prediction is increased. In particular, the variance of each regression coefficient in A is increased. This can be illustrated by choosing $\mathbf{x}$ as a vector of zeros with a one in the pth place, $p \leq a$, and $\mathbf{x_A}$ identical to $\mathbf{x}$ except that the last $\mathbf{b}$ elements in $\mathbf{x}$ are eliminated. Such a vector merely selects the pth element of $\hat{\beta}$ as well as $\hat{\beta}_A$. However, including too few predictors results in what is known as omission bias. Suppose now that at least one predictor in set B is nonredundant. Having only selected the predictors in set A we could calculate $E(\hat{\beta}_A)$ as

$$E(\hat{\beta}_A) = \beta_A + (\mathbf{X'_A X_A})^{-1} \mathbf{X'_A X_B} \beta_B,$$

which results in biased prediction. The second term in the above equation gives the amount of shift between the true value of β_A and the expected value of its estimator. The bias, that is, the difference between the true value and the expected value of our prediction, is

$$\text{bias}(\hat{\mathbf{y}}_A) = \text{bias}(x'_A \hat{\beta}_A) = \mathbf{x'_A} - \mathbf{x'_A} (\mathbf{X'_A X_A})^{-1} \mathbf{X'_A X_B} \beta_B.$$

A derivation of these results can be found in Miller (1990). A bias in an estimator is not a problem as long as it is not 'too' large. For example, the usual estimate of the standard error s where, $s^2 = \frac{1}{n-1} \sum_{i=1}^{n} (x_i - \bar{x})^2$, is biased as well. For an unbiased estimator of the standard deviation see, for example, (Arnold, 1990, p. 266). (However, s^2 is an unbiased estimator for the population variance.) There are techniques available for detecting and reducing the omission bias (Miller, 1990).

The crucial point here is to see that the aim of variable selection lies in selecting just enough variables so that the omission bias is small and, at the same time, increasing the variance of the prediction or, equivalently, of the regression coefficients not more than necessary. In Miller's words, "we are trading off reduced bias against increased variance" (1990, p. 6).

A few words about the variables that enter into the regression equation are in order. Of course, the dependent variable should, perhaps after a suitably chosen transformation, be approximately normally distributed. If there is evidence that the relation between the dependent and an independent variable is curved, quadratic or even higher order terms should be included in the set of all predictors. The same holds true for any interaction terms, like $x_p x_q$, between any predictors. Note also that it is usually not meaningful to include, for example, a quadratic term, say x_p^2, in the equation without x_p or an interaction term without at least one of the variables contained in the interaction. All of the techniques from regression diagnostics can be used to make sure that the usual regression assumptions are at least approximately satisfied. It should be noted that this examination is, although recommended, incomplete, because the final, and hopefully adequate, regression model is not yet determined. Model checking is very useful after a final model has been selected.

Further, it should be noted that a regression model can only be fit to data if there are at least as many observations as there are predictors. If we have more predictors than observations some of the variable selection techniques discussed in the following will not work, for instance, backward elimination. Often researchers wish to include a variable merely on theoretical, rather than statistical, grounds. Or, it may make sense to include or exclude an entire set of variables, as is often the case with dummy variables used for coding a factor. These options should be kept in mind when interpreting results of a variable selection technique.

Variable selection techniques can be divided into "cheap" ones and others. The first group enters or removes variables only one at a time. Therefore they can be performed with large numbers of independent variables. However, they often miss good subsets of predictors. On the other hand, the other techniques virtually guarantee to find the "best" subsets for each number of predictors but can be performed only if the number of predictors is not too large. We discuss both groups in different sections, starting with the best subset regression technique. But first, we present an example in order to be able to illustrate the formulas given below.

13.1   A Data Example

The data are taken from a study by von Eye et al. (1996) which investigated the dependence of recall on a number of cognitive as well as demographic variables. We use a subset of the data. This subset contains 183 observations, 10 predictors, and the dependent variable. The data are given in Appendix E.1. Each subject in the study was required to read two texts. Recall performance for each text was measured as the number of correctly recalled text propositions. The two recall measures were added to yield a single performance measure. There were two types of text in the experiment, concrete texts and abstract texts. The texts had

Table 13.1: *List of Variables Used to Illustrate Best Subset Selection Techniques*

1. AGE	subject age in years	
2. EG1	dummy variable for experimental group	
3. SEX	dummy variable for sex	
4. HEALTH	4-level rating scale from very good to poor	
5. READ	reading habits in hours per week	
6. EDUC	7-level rating scale indicating highest degree of formal schooling completed	
7. CC1	measure of cognitive complexity: breadth of concepts	
8. CC2	measure of cognitive complexity: depth of concepts	
9. OVC	measure of cognitive overlap of concepts	
10. TG	dummy variable for type of text: abstract vs concrete	
11. REC	dependent variable recall performance	

been created to tap cohort-specific memories. For instance, it is assumed that cohorts have good memories of music fashionable when cohort members were in their teens. Later, many cohort members spend less time listening to music. Therefore, the music of their teens stays prominently with them as cohort-specific memory content. The same applies to such abstract concepts as heros and educational goals, and to such concrete concepts as clothing styles. Consider the following example.

Individuals that were middle-aged around 1985 grew up listening to Elvis Presley music. Individuals twenty years older grew up listening to Frank Sinatra music, and individuals twenty years younger grew up listening to Bruce Springsteen music. The texts had been constructed to be identical in grammatical structure and length, but differed in the name of musician mentioned. This was done in an analogous fashion for the other topics. The hypothesis for this part of the experiment was that at least a part of the ubiquitous age differences in memory performance can be explained when differential, cohort-specific familiarity with contents is considered.

The 10 explanatory variables and the dependent variable *REC* are given in Table 13.1.

First of all, we draw a histogram of the dependent variable which

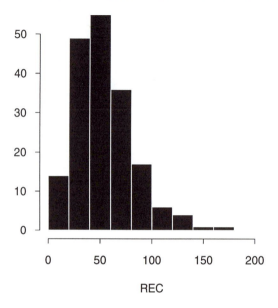

Figure 13.1: Histogram of Recall, REC, in cohort memory experiment.

is shown in Figure 13.1. This plot shows clearly some skewness in the distribution of the variable *REC*.

The normal probability plot, Figure 13.2, shows a clear departure from normality.

Therefore, a transformation of *REC* may be worthwhile. As this variable is a number of counts and counts often follow a Poisson distribution a square root transformation is recommended to stabilize the variance and to obtain an approximately normally distributed variable. Let Y be defined as

$$Y = \sqrt{\text{REC}}.$$

After this transformation, the normal probability plot for Y now shows no systematic departures of Y from normality (see Figure 13.3). For the following analyses we use Y as the dependent variable.

All scatterplots of each potential explanatory variable against Y (not shown here) show wide scatters with no curvature that would require higher order terms of the explanatory variables. No correlation between

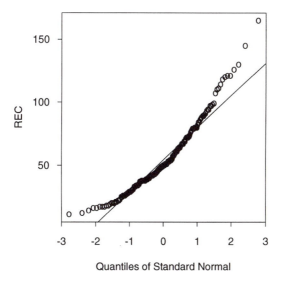

Figure 13.2: Normal probability plot of raw frequencies of Recall.

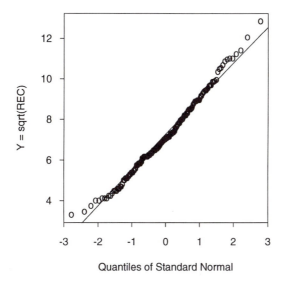

Figure 13.3: Normal probability plot of Recall rates after square root transformation.

Y and one of the continuous predictors is higher in absolute value than $r = 0.25$. For the three dummy variables in the data set the means of Y are calculated for the two groups belonging to each dummy variable, showing a considerable shift in the mean for TG (8.04 vs. 6.36) and $EG1$ (8.68 vs. 6.84) but only a small difference for SEX (7.24 vs. 7.12). From this information we should probably select TG and $EG1$ as predictors but it is not clear which of the continuous predictors to select in order to improve the model fit. Now we will show how variable selection techniques can be used as a guide to answer this question.

13.2 Best Subset Regression

In principle, best subset regression is straightforward. One "merely" has to compute the regression equation for each possible subset of the, say k, available predictors and then use some goodness-of-fit criterion, for instance, R^2, to decide which set of predictors yields a good or possibly best fit. Before discussing how "best" could be defined we should note how many possible subsets can be produced from k predictors. The answer is 2^k as there are $\binom{k}{0} = 1$ possible ways to select no predictor (fitting only the intercept term), $\binom{k}{1} = k$ possible ways to select one predictor, $\binom{k}{2} = k(k-1)/2$ ways to select two predictors, and so on. If one sums all the possibilities, that is,

$$\sum_{i=0}^{k} \binom{k}{i},$$

the result is 2^k. With 10 predictors there are 1024 possible regression models and we could easily spend a whole day fitting all the models, and with 20 predictors fitting all models will take (at least) a whole lifetime, 1,048,576 regression runs. Thus, speedy algorithms are needed to reduce the amount of work.

There are two types of algorithms; the first type calculates all possible regressions (see Schatzoff, Tsao, & Fienberg, 1968) whereas the second type only gives a few of the best subsets for each number of independent variables (see Furnival & Wilson, 1974). There has been considerable work in this area and today some of these algorithms are implemented in standard computer packages, for example, SAS and S-Plus. (See reference manuals for details.) How these algorithms work need not concern us here. We next explain how to judge the fit of a regression model.

There are several criteria in use for deciding how well a regression model fits the data. Therefore, when speaking of a "best" subset of predictors this always refers to the optimality criterion chosen. We discuss the following criteria: R^2, adjusted R^2, and Mallow's C_p.

13.2.1 Squared Multiple Correlation, R^2

First recall the definition of R^2, the square of the multiple correlation coefficient,

$$R^2 = \frac{\text{SSReg}}{\text{SSR}_m}.$$

This is just the sum of squares that can be explained by the regression model, SSReg, divided by the sum of squares that can be explained when only the intercept is fitted, SSR_m; that is, no sample information is used to explain the dependent variable. (SSR_m is also known as the total sum of squares corrected for the mean.) As SSR_m is a constant for given data, R^2 and SSReg as well as $\text{SSRes} = \text{SSR}_m - \text{SSReg}$ are entirely equivalent ways of expressing model fit. Maximizing R^2 corresponds to minimizing SSRes, the residual sum of squares. All the following explanations are given in terms of R^2.

It could be argued that the higher R^2, the better the fit of the model. But this argument leads directly to selecting the model with all the predictors included, as R^2 could never decrease with the addition of a further predictor. Formally, this is

$$R^2(\text{Model I}) \leq R^2(\text{Model II})$$

if Model II contains all the predictors of Model I plus one additional predictor. For this reason, we do not concentrate on the highest R^2 value but on large size changes in R^2 between models with different numbers of predictors. A program for best subset selection will print out the model or perhaps a few models with the highest R^2 values for each number of independent variables if the R^2 measure is selected as the goodness-of-fit criterion. From this output it can be seen when further addition of predictors will not increase R^2 considerably. If one is after explanation rather than prediction one should always request (if possible) a few of the best models for each number of predictors, as there are usually some models with nearly identical fit, in particular if the number of predictors is high.

To save space, Table 13.2 gives only the three best subsets of one to six predictors (taken from Table 13.1) in the equation according to the R^2 criterion for the present sample data. If we enter all 10 predictors in the equation we have $R^2 = 45.16\%$, that is, we can explain about half of the variation of Y by our explanatory variables. It is hard to say whether this indicates that some important variables have not been recorded since a relatively low R^2 does not imply a poor model fit. The low R^2 was expected anyway as the correlations found between the continuous predictors and Y are very low. One can see from the table the outstanding role of the variables TG (10) and $EG1$ (2). TG accounts for 21% and $EG1$ for 16% of the variation in Y, and both variances add together to 37% if the model is fitted with both variables in the equation.

When there are three or more variables in the equation, all models in the table contain TG as well as $EG1$. First, this shows that the experimental condition applied had a considerable effect. Also as expected, text group (abstract vs. concrete) has a high influence on recall.

Looking at the models with three predictors we see that the cognitive measures enter the model next but the increase in R^2 is relatively low and the model fits are very similar. The correlations of OVC with $CC1$ and $CC2$ are -0.78 and -0.54, respectively, indicating that there is some overlap between these measures. The correlation between $CC1$ and $CC2$ is, although significant at the 5% level, virtually zero, $r = 0.15$. As the difference between the best model with three variables and the best model with five or six variables in the equation is only about 3 OVC is a reasonable choice, although prediction could be slightly improved when more variables are added. Note also that from three variables in the equation onward there are no big differences between the model fits. Thus, theoretical considerations may have to guide researchers when selecting one of the other models.

In some sense the other two measures, the adjusted R^2 and Mallow's C_p, penalize the addition of further variables to the model, that is, with

Table 13.2: *Best Three Subsets for One to Six Predictors According to the R^2, Adj R^2, and C_p criteria. R^2 and Adj R^2 are given in percent. The first column gives the number of predictors in the equation in addition to the intercept.*

No. of Var.	Var. in the Equation[a]	R^2	Adj R^2	C_p
1	(10)	20.99	20.55	68.83
	(2)	15.92	15.46	84.72
	(9)	5.45	4.93	117.56
2	(2,10)	37.01	36.31	20.58
	(5,10)	25.08	24.25	57.98
	(9,10)	24.20	23.36	60.76
3	(2,9,10)	41.26	40.27	9.25
	(2,8,10)	40.35	39.35	12.11
	(2,7,10)	39.14	38.12	15.91
4	(1,2,9,10)	42.71	41.42	6.69
	(2,5,9,10)	42.35	41.05	7.83
	(2,8,9,10)	41.97	40.66	9.03
5	(1,2,5,9,10)	44.19	42.61	4.05
	(1,2,6,9,10)	43.64	42.05	5.78
	(1,2,7,8,10)	43.33	41.72	6.76
6	(1,2,5,8,9,10)	44.59	42.70	4.80
	(1,2,5,6,9,10)	44.56	42.67	4.90
	(1,2,5,7,8,10)	44.51	42.62	5.06

[a]From Table 13.1

these criteria the model fit can decrease when adding variables to the model.

13.2.2   Adjusted Squared Multiple Correlation

The adjusted R^2 is defined as

$$Adj\ R^2 = 1 - \frac{\text{MSRes}}{s_y^2},$$

where s_y^2 is just the sample variance of the dependent variable. As this is a constant, for a given set of data, Adj R^2 and MSRes are again equivalent ways for expressing this criterion. Note that Adj R^2 accounts for the number of predictors through MSRes, as this is just $SSRes/(n-p)$, where n is the number of observations and p the number of predictors in the equation, including the intercept. If a predictor is added that explains nothing, it could not decrease SSRes, but the calculation of MSRes has lost one degree of freedom, that is, the denominator of MSRes is reduced by one. Thus, MSRes increases. The model with the highest *Adj R^2* should therefore be judged as the best model regardless of the number of predictors in the model. But this statement should be taken with a grain of salt, as it can be shown that the *Adj R^2* criterion used in this strict sense has the tendency to include too many variables.

The strategy in model selection should therefore be the same as with the R^2 criterion. Using a computer program obtain a few of the best subsets for each distinct number of predictors and select a model considering (1) the relative increase of *Adj R^2* as more variables are entered into the model and (2) theoretical arguments.

The second to last column of Table 13.2 shows the best three subsets of one to six variables in the model using the *Adj R^2* criterion. Indeed, it is no accident that the best three subsets for a given value of p are the same whether the R^2 or the *Adj R^2* criterion is used. This also generalizes to the C_p criterion, to be discussed below, as it can be shown that, for a given value of p, the R^2, *Adj R^2*, and C_p criteria all induce the same ranking of the models. Formally this is

$$
\begin{aligned}
R_p^2(\text{Model I}) &\leq R_p^2(\text{Model II}) \\
Adj\ R_p^2(\text{Model I}) &\leq Adj\ R_p^2(\text{Model II}) \\
C_p(\text{Model I}) &\geq C_p(\text{Model II}),
\end{aligned}
$$

where Model I and Model II both contain a subset of p variables but the two subsets are different. By looking at Table 13.2 we come to the same conclusions as before. It is of interest that whereas the *Adj R^2* criterion penalizes the addition of further variables, *Adj R^2* is slightly increased when adding a fourth and a fifth variable to the model. But recall that the *Adj R^2* criterion has the tendency to include more variables into the model than actually needed.

13.2.3 Mallow's C_p

The derivation of Mallow's C_p can be found, for example, in Christensen (1996) or Neter et al. (1996). It is assumed that the model with all predictors is the correct model and thus we can estimate the true residual variance σ^2 by

$$\hat{\sigma}^2 = \frac{\text{SSRes}(k)}{n-k}.$$

Recall that k is the number of available predictors and n denotes the number of observations. SSRes(k) is the residual sum of squares with all the predictors in the model and by assumption the true model as well. Let SSRes(p) be the residual sum of squares with only p of the k predictors in the model. Mallow's C_p is then given as

$$C_p = \frac{\text{SSRes}(p)}{\hat{\sigma}^2} - (n - 2p).$$

To better understand Mallow's C_p consider that, from a statistical viewpoint, it is desirable to minimize the expression

$$\frac{1}{\sigma^2} E(\hat{\mathbf{y}}(p) - \boldsymbol{\mu})'(\hat{\mathbf{y}}(p) - \boldsymbol{\mu}),$$

where σ^2, the residual variance, is just a scale parameter, $\hat{\mathbf{y}}(p)$ denotes the prediction using only p predictors, and $\boldsymbol{\mu}$ denotes the true but unknown mean response. Thus we evaluate the fit of a regression model by looking at the expected squared distance between the true value and the prediction given from some model. The formula is an expected value of a quadratic form involving population parameters and thus it is typically unknown to us. Mallow's C_p is an estimator of this expression. If p predictors are sufficient to provide a good description of our data, then Mallow's C_p is as small as the distance between $\hat{\mathbf{y}}(p)$ and $\boldsymbol{\mu}$ is small. For this reason we are interested in finding regression models with small C_p values. If a subset of p predictors can explain the dependent variable very well, the expected value of C_p can be shown to be

$$E(C_p) = p\frac{2(k-p)}{n-k-2}.$$

If the sample is large relative to the number of predictors needed for a good description of the data, that is, $n \gg k$ and p, the second term in the above equation will be small, as n is in the denominator, and $E(C_p) \approx p$. Hence, a good model yields a small C_p value that is near p.

The C_p values for our example are again given in Table 13.2 for the three best subsets from one to six predictors. First note that the differences in C_p within each subset of predictors are larger than the differences of the other criteria. If one is willing to select a model with only three predictors in it, C_p suggests using the OVC variable instead of the other two cognitive measures, $CC1$ and $CC2$. With C_p, we should select models for which $C_p \approx p$, so a model with only three variables in the equation may not be good enough. After five variables are entered, that is, $p = 6$ (remember the intercept), C_p is close to six, so each of the models given in the table may be a reasonable choice. These models contain at least one of the three cognitive complexity variables. With six variables entered into the equation, C_p increases slightly compared to the best model with only five variables in the equation and the value of C_p is slightly too low. Thus, according to C_p, we select one of the five-variable models.

So, what we have obtained from best subset regression is not "the" best model, but we have identified a few good models, which leaves it to us to decide on theoretical grounds which model we finally select. We should now investigate with regression diagnostic techniques whether the variance of the residuals is approximately constant, whether the residuals are normally distributed, whether there are influential observations, and so on.

For illustrative purposes we select the third best model with five predictors plus the intercept. For this model we calculate $C_p = 6.76$. This is a relatively parsimonious model because apart from the indicators for text group and experimental condition it contains $CC1$ and $CC2$, the two nearly uncorrelated measures of Depth and Breadth of Cognitive Complexity, and Age, which is reasonable as the experimental conditions varied the cohort-specific content of the texts. From the plot of the standardized residuals against the fitted values in Figure 13.4 the variance looks reasonably stable. While the bulk of the observations lie in the interval between -2 and 2, there is at least one extreme residual with a value of about 4. With this observation something unusual has happened. The model had predicted a low value for Recall but the observation has a high

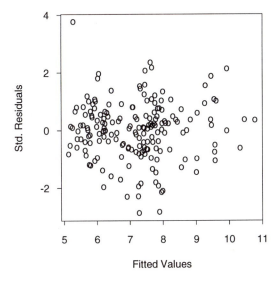

Figure 13.4: Plot of fitted values against standardized residuals in cohort memory experiment.

positive residual, indicating that this person performed much better than expected from his or her predictor values.

The normal probability plot given in Figure 13.5 shows that the standardized residuals are approximately normally distributed, but the tails of the distribution are heavier than those of a normal distribution.

While best subset regression can be easily determined if the number of predictors is not too large, say, less than 40, other methods are needed that are computationally less intensive if the number of predictors is considerably higher. For this reason, these are usually referred to as the "cheap" methods.

13.3 Stepwise Regression

There are three different methods that enter or delete variables to or from the model one at a time. These are forward selection, backward elimination, and the Efroymson algorithm. Often the name "stepwise regression" is used for the algorithm proposed by Efroymson (1960).

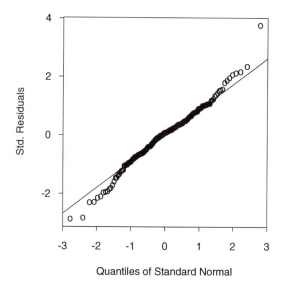

Figure 13.5: Normal probability plot in cohort memory experiment.

13.3.1   Forward Selection

The idea behind forward selection is quite simple. Having k predictors we calculate k simple linear regression models, one for each predictor. The model with the highest F value should be a reasonable choice for selecting a variable at this stage of the variable selection process. Recall that the F value for a simple linear regression is defined as

$$F_1(X_r) = \frac{\mathrm{MSReg}(X_r)}{\mathrm{MSRes}(X_r)} = \frac{\mathrm{SSReg}(X_r)/1}{\mathrm{MSRes}(X_r)},$$

where the subscript 1 on F means that this is the F value for deciding upon the first variable, and $MSRes(X_r)$ is the residual mean square with only the rth predictor (and of course the intercept) in the equation, $MSReg(X_r)$ denotes the sum of squares accounted for by the rth variable divided by its degree of freedom. Since the numerator $df = 1, \mathrm{MSRes}(X_r) = \mathrm{SSReg}(X_r)$. The variable with the highest F_1 value is the variable selected by this procedure. Now, this procedure is repeated where the variables selected at earlier stages are always part of the model, that is, once a variable is selected by forward selection it remains in the equation until the procedure stops. To repeat the procedure means each of the $k - 1$ remaining variables, for instance, X_s, is selected and entered into the equation and the increase in SSReg is observed, that is, $\text{SSReg}(X_r, X_s) - \text{SSReg}(X_r)$, and related to $\text{MSRes}(X_r, X_s)$. This means that for each of the remaining variables F_2 is now calculated as

$$F_2(X_s | X_r) = \frac{\text{SSReg}(X_r, X_s) - \text{SSReg}(X_r)}{\text{MSRes}(X_r, X_s)}.$$

The variable with the highest F_2 value is selected by the procedure. Again, this variable stays in the model until the variable selection process terminates. Having selected two variables we select the third out of the remaining $k - 2$ by finding that variable, say, X_t, with the largest F_3, where

$$F_3(X_t | X_r, X_s) = \frac{\text{SSReg}(X_r, X_s, X_t) - \text{SSReg}(X_r, X_s)}{\text{MSRes}(X_r, X_s, X_t)}.$$

Of course, the procedure needs a stopping rule that terminates the variable selection process if none of the remaining variables can improve the model fit considerably. Typically, the highest F value is compared to a predetermined value which can usually be specified as an option in computer programs. This is known as the *F-to-enter* test. As long as the maximal F value calculated at any stage is higher than the critical value, a variable is selected. If no variable can fulfill this condition the procedure stops.

There are a few equivalent criteria for terminating forward selection. These include the highest partial correlation, the highest increase in R^2, or the highest t statistics for the coefficients instead of the highest F values deciding upon which variable, if any, should be entered next. These procedures all yield the same results – assuming the termination criteria are accordingly modified.

Before applying the forward selection method to our sample data we have to specify a critical F value. As we have 183 observations, the F value for a regular F test at the 5% level would be about $F(0.05, 1, 183) = 3.9$. Thus, for ease of presentation we take a critical value of $F_c = 4.0$ to decide whether a variable should be entered or not. Table 13.3 gives all

Table 13.3: *Results from the Forward Selection Procedure*

Step	Var. Entered	SSReg	MSRes	F
1	TG	128.67	2.67	48.08
2	EG1	226.89	2.14	45.89
3	OVC	252.94	2.01	12.96
4	AGE	261.84	1.97	4.52
5	READ	270.91	1.93	4.70

the relevant information.

Having entered these five variables, the next highest obtainable F value is 1.26, for variable $CC2$. The F values in the table can easily be recalculated. For instance, for variable OVC we have $F = (252.94 - 226.89)/2.01 = 12.96$. The result of this variable selection corresponds to the best subsets containing one to five variables, as can be seen from Table 13.3. The model selected is one of the most promising obtained by best subset selection. Indeed this is the model with the lowest C_p value. But it should also be noted that because the procedure ends with one final model we get no information that there are other models with virtually the same fit. This is also true for the other two stepwise procedures which are discussed below.

13.3.2 Backward Elimination

Backward elimination is the opposite of forward selection. It starts with the full model, that is, all variables are entered in the equation (assuming that there are more observations than variables). A reason for a variable, say X_r, to be eliminated is that there is only a small loss in model fit after that variable is removed. Hence, the F_k value, that is, the test whether X_r should be removed if there are all k variables in the model, is small if X_r adds close to nothing to the model. For each of the variables in the full model F_k is calculated and the variable with the smallest F_k value is removed from the model. Formally, F_k is

$$F_k(X_r) = \frac{\mathrm{SSReg}(X_1,\dots,X_k) - \mathrm{SSReg}(X_1,\dots,X_{r-1},X_{r+1},\dots,X_k)}{\mathrm{MSRes}(X_1,\dots,X_k)}.$$

The procedure is repeated after elimination of the first variable. Once a variable is removed from the model this variable is not reevaluated to see whether it could possibly at a later stage improve the model fit. Having deleted, say, X_r, we check whether in the reduced model other variables should be deleted by calculating F_{k-1},

$$F_{k-1}(X_s|X_r) = \frac{\text{SSReg}(X_1,\dots,X_{r-1},X_{r+1},\dots,X_k)}{\text{MSRes}(X_1,\dots,X_{r-1},X_{r+1},\dots,X_k)}$$
$$- \frac{\text{SSReg}(X_1,\dots,X_{r-1},X_{r+1},\dots,X_{s-1},X_{s+1},\dots,X_k)}{\text{MSRes}(X_1,\dots,X_{r-1},X_{r+1},\dots,X_k)}.$$

While the notation gets rapidly messy, the idea behind it is quite simple. Again, we need a stopping rule and, as before, a critical F value is chosen and the minimal empirical F value at any stage is compared to the critical one. This comparison is known as the *F-to-delete* or *F-to-remove* test. As long as we can find variables satisfying this condition the procedure continues to eliminate variables and stops otherwise. The critical F value can usually be specified by the user of a computer program.

Now we apply the backward elimination procedure to our data. Again, we use 4.0 as our critical F value, meaning that we remove variables from the full model as long as we can find a variable with a smaller F value than 4.0. The results are given in Table 13.4.

The F values are again easily obtained. For example, in step four we calculate $F = (275.50 - 273.35)/1.92 = 1.120$. For the first F value we calculate $F = (276.88 - 276.59)/1.95 = 0.149$. After these five variables have been removed from the full model a further removal of any variable

Table 13.4: *Results of the Backward Elimination Procedure. In the full model* $SSReg = 276.88$, $MSRes = 1.95$

Step	Var. Removed	SSReg	MSRes	F
1	CC1	276.59	1.94	0.149
2	SEX	276.06	1.94	0.273
3	HEALTH	275.50	1.92	0.289
4	EDUC	273.35	1.93	1.120
5	CC2	270.91	1.93	1.264

will raise the F value over the critical value of 4.0, so we end up with the same model as selected by forward elimination. This is a lucky situation as the procedures usually give different final models. All remarks concerning the selected model for forward selection apply here .

13.3.3   Efroymson's Algorithm

Efroymson's algorithm is a combination of forward selection and backward elimination. An obvious disadvantage of the stepwise methods discussed so far is the inability to correct decisions once made by the procedures, that is, to remove a variable once selected with forward selection or to re-enter a variable once removed with backward elimination. The algorithm suggested by Efroymson overcomes this disadvantage. As in forward selection the procedure starts by entering the variable which yields the highest F_1 value and satisfies the *F-to-enter* test. Likewise, the second variable is selected. But now the procedure checks whether any of the two variables entered so far can be removed with the *F-to-delete* test. Generally, after a variable is selected (other than the first) with the *F-to-enter* test the procedure tries to eliminate variables that may no longer be needed for good model fit using the *F-to-delete* test. Therefore, computer programs allow for user defined values of two critical F values, one for the *F-to-enter* test and one for the *F-to-delete* test.

Application of Efroymson's algorithm to our data yields the same table as the one for forward selection, so it is not given here. The situation where one of the previously selected variables has become obsolete does not occur with these data.

A few words about which values to select for the critical F values are in order. First of all, it is stressed again that the calculated F values do not follow an F distribution under the null hypothesis, which states that the model under test is actually the true model, taking for granted that the usual assumptions like independence, normality of the residuals etc., hold. In addition, there is the problem of multiple testing. Also, even without these problems, it would be unreasonable to compare the calculated F value to a single fixed critical value, because the degrees of freedom for MSRes in the denominator of the above formula change as variables are entered or removed from the model. (Note that the degrees of freedom in the numerator remain one.) Both the *F-to-enter* and the

F-to-delete test are of heuristic value only and should not be interpreted in a probabilistic manner. Typically computer programs provide default values for the critical F values, for example, 2.0 for the *F-to-enter* value. The idea behind this value is that the sum of squares accounted for by a variable should be at least twice as large as the corresponding mean squared error. Another option is to select critical F values guided by the F distribution. If, for instance, 50 observations are available, a critical F value of 4 might be appropriate since F(0.95;1;50) = 4.03. The F value for the *F-to-delete* test should always be less than the value for the *F-to-enter* test. Otherwise, Efroymson's procedure might not come to an end.

13.4   Discussion

Variable selection techniques should be considered as tools of exploratory data analysis (EDA) (Tukey, 1977) since hypothesis tests and confidence intervals obtained from the same data on which the variable selection process was based are invalidated because of the bias in the estimates. Typically the models fit the data better than they fit in the population. This can be explained as follows. If one imagines a replication study, the parameter estimates for the same model as selected from the first data set will almost certainly be smaller for the replication data. If there are many observations available it is possible to overcome this problem by dividing all observations into two different sets. The first set is then used for model selection, and parameter estimates are obtained from the second set.

Because stepwise regression techniques end up with a final model, it is tempting to think of the final model as "the" best model, which could be far from being true. As can be seen from best subset regression there is usually a number of models with a nearly identical fit. As long as the amount of data permits one to use best subset regression techniques it should be done. This is the case if there are not considerably more than 40 variables available. With more than 40 variables the stepwise procedures must be used.

Another problem with variable selection has to do with influential observations. Recall that observations might appear to be influential because one or a few important variables have been omitted from the model.

If these variables are added the observations are possibly no longer influential. Selecting a smaller model and deleting the influential observations can alter the model fit considerably because the deleted observations are by definition influential.

As long as the problems of variable selection techniques are kept in mind and they are seen as tools for doing EDA, they can give insights into the subject at hand, especially if no alternative data analysis procedure can be recommended.

Chapter 14

REGRESSION FOR LONGITUDINAL DATA

There are situations where we have not only one measurement of a personal characteristic, but we have repeatedly observed a sample of individuals over time. Thus, we have several measurements for each individual. While it is usually reasonable to assume that measurements made on different individuals are independent and hence uncorrelated, this assumption is generally not accepted if an individual is measured on the same characteristic several times. The measurements within an individual are usually positively correlated over time. Often one observes that the correlation decreases as the time interval between measurements increases; that is, measurements made close together in time are more related to each other than measurements farther apart. But there are situations where it would be more sensible to think of the correlations between measurements within a person as constant. We will deal with this topic later. While measurements within a person are usually correlated, those between individuals are thought of as being independent. Note that in this book we are discussing models for normally distributed data so the assumption of correlation between observations is equivalent to the assumption of dependent measurements. When we are interested in relating observations to other (independent or explanatory) variables, the familiar regression approach can be used.

Thus, in longitudinal data analysis the goal of analysis is the same as throughout the whole book, that is, to identify those explanatory variables that can explain the dependent variable and to estimate the magnitude of the effect each independent variable has. The analysis is complicated by the correlated observations within a person. This characteristic of repeated measures regression can be seen in contrast to two other well-known techniques for data analysis. In time series analysis we usually have only one or possibly a few long series of observations and the focus is on describing the relatedness of the observations to each other over time. In multivariate statistics there is a sample of individuals and for each individual we measure several variables. These observations per individual are typically not thought of as being related over time but cover various aspects of the individuals, e.g., miscellaneous mental abilities. The interdependence of the data for each individual is, therefore, not as highly structured as in longitudinal data analysis.

Approaches to the analysis of longitudinal data differ from each other, among other aspects, in the allowable time pattern with which data are collected. If the repeated observations are made for each individual at the same point in time or the intervals between repeated observations are the same for all individuals, analysis is considerably simplified. In the following we describe how to analyze longitudinal data when the measurements are made at the same points in time for each individual but for the intervals between measurements are not necssarily equal. For an overview of approaches to analyzing longitudinal data, see Ware (1985).

14.1   Within Subject Correlation

Let us first consider the measurements made on one individual. Let y_{ij} be the observation on the ith individual, $i = 1, \ldots, m$, made at the jth point in time, $j = 1, \ldots, n$. Note that this notation implies that the pattern of observation points in time is the same for every individual. Associated with each y_{ij} is a $(p \times 1)$ vector of explanatory variables $\mathbf{x}'_{ij} = (x_{ij1}, x_{ij2}, \ldots, x_{ijp})'$ and we assume that y_{ij} could be written as a linear model,

$$y_{ij} = \beta_0 + \beta_1 x_{ij1} + \beta_2 x_{ij2} + \cdots + \beta_p x_{ijp} + \varepsilon_{ij}.$$

The ε_{ij} is the residual error term, which is now assumed to be correlated within an individual, that is, $\rho = corr(\varepsilon_{ij}, \varepsilon_{ij'})$ for $j \neq j'$. All n measurements of one individual can be expressed using matrix notation. For individual i one writes

$$\mathbf{y_i} = \mathbf{X_i}\boldsymbol{\beta},$$

where $\mathbf{y_i'} = (\mathbf{y_{i1}}, \mathbf{y_{i2}}, \ldots, \mathbf{y_{in}})'$, $\mathbf{X_i}$ is the $(n \times p)$ design matrix for individual i, and $\boldsymbol{\beta'} = (\beta_1, \beta_2, \ldots, \beta_p)$ is the $(p \times 1)$ vector of unknown regression weights. The dependence of the n measurements of individual i is represented by the covariance matrix $\boldsymbol{\Sigma_s}$, which has no subscript i because the covariance matrix is assumed to be the same for each individual. The subscript s indicates that this is the covariance matrix for a single person to separate it from the covariance matrix of all individuals defined below. This is just a generalization of the assumption of constant variance in the preceding chapters. Notice also that the explanatory variables are generally allowed to vary over time although the analysis is further simplified if they do not. The notation for the complete set of $N = mn$ observations is quite similar. One just drops the subscript for individual i. Let $\mathbf{y'} = (\mathbf{y_1'}, \ldots, \mathbf{y_m'})$ be the $(mn \times 1)$ vector of observations and $\mathbf{X'} = (\mathbf{X_1'}, \ldots, \mathbf{X_m'})$ the $(mn \times p)$ design matrix. Then, one writes

$$\mathbf{y} = \mathbf{X}\boldsymbol{\beta}.$$

The covariance matrix for all observations has block diagonal form. Specifically, it is

$$\Sigma = \begin{pmatrix} \Sigma_s & 0 & \cdots & 0 \\ 0 & \Sigma_s & \cdots & 0 \\ \vdots & \vdots & \ddots & \vdots \\ 0 & 0 & \cdots & \Sigma_s \end{pmatrix}.$$

It is very important to think about possible processes that occur in within subject correlation. One possibility that may occur in uniform correlations within a subject is well known from the analysis of variance. Suppose that subjects are measured repeatedly over time in a designed experiment. In a designed experiment the values of the covariates are the same for each observation in a treatment group. For example, the productivity of workers is observed under three different illumination conditions. Such an experiment is usually referred to a repeated measurement design. Of course, this can also be seen as a mixed model, treating the illuminating conditions as a fixed factor and the workers as the random factor. The model used in the analysis of variance is usually

$$y_{ij} = \mu_j + \tau_i + \varepsilon_{ij},$$

where μ_j represents the effect of the illumination conditions on y. ε_{ij} are the residual error terms which are supposed to be independently normally distributed with constant variance, that is, $\varepsilon_{ij} \sim N(0, \sigma_\varepsilon^2)$, and the τ_i are also independent random variables, normally distributed, $\tau_i \sim N(0, \sigma_\tau^2)$. The τ_i index the workers. Both random terms are assumed to be mutually independent. Note that $\varepsilon_{ij} = \tau_i + \epsilon_{ij}$, where ε_{ij} are the correlated errors in the linear model formula from above. In the example, the random term for workers means that some workers are more productive than others, varying with τ_i, and that this is independent of errors under the different illuminating conditions. Generally speaking, there are high scorers and low scorers. We now derive the variance of a single observation under the above model and given assumptions. $Var(y_{ij}) = Var(\mu_j + \tau_i + \epsilon_{ij}) = Var(\tau_i) + Var(\epsilon_{ij}) = \sigma_\tau^2 + \sigma_\epsilon^2$. For this derivation it is essential that the two random terms be independent of each other. Next we look at the covariance of two observations within a subject, i.e., $Cov(y_{ij}, y_{ij'})$ for $j \neq j'$,

$$
\begin{aligned}
Cov(y_{ij}, y_{ij'}) &= E(y_{ij} - \mu_j)(y_{ij'} - \mu_{j'}) \\
&= E(t_j + \epsilon_{ij})(\tau_i + \epsilon_{ij'}) \\
&= E(\tau_i^2 + \tau_i \epsilon_{ij} + \tau_i \epsilon_{ij'} + \epsilon_{ij} \epsilon_{ij'}) \\
&= E(\tau_i^2) + E(\tau_i \epsilon_{ij}) + E(\tau_i \epsilon_{ij'}) + E(\epsilon_{ij} \epsilon_{ij'}) \\
&= E(\tau_i^2) \\
&= \sigma_\tau^2.
\end{aligned}
$$

We see that under this model all within subject correlations are equal. This is an essential assumption for the above analysis of variance models. The covariance matrix of the repeated observations of a single subject Σ_s has therefore the following form:

$$\Sigma_s = \begin{pmatrix} \sigma_\epsilon^2 + \sigma_\tau^2 & \sigma_\tau^2 & \cdots & \sigma_\tau^2 \\ \sigma_\tau^2 & \sigma_\epsilon^2 + \sigma_\tau^2 & \cdots & \sigma_\tau^2 \\ \vdots & \vdots & \ddots & \vdots \\ \sigma_\tau^2 & \sigma_\tau^2 & \cdots & \sigma_\epsilon^2 + \sigma_\tau^2 \end{pmatrix}.$$

The correlation matrix, $\mathbf{R_s}$, is obtained from Σ_s by dividing each element by $\sigma_y^2 = \sigma_\epsilon^2 + \sigma_\tau^2$.

$$\mathbf{R_s} = \begin{pmatrix} 1 & \rho & \cdots & \rho \\ \rho & 1 & \cdots & \rho \\ \vdots & \vdots & \ddots & \vdots \\ \rho & \rho & \cdots & 1 \end{pmatrix}.$$

It follows that observations on different subjects have zero covariance so that the correlation matrix of all observations is of block diagonal form with uniform correlations as off-diagonal elements in each block. The uniform correlation model supposes that the correlation between observations does not decay with time. This situation is quite different for another prominent way of building correlation into repeated observations. This way is known as the first-order autoregressive process, which is also known as the first-order Markov process. Here the correlations decay with time. This model is only appealing for equidistant observations. The observation of individual i at time point j, y_{ij}, is thought of as having a fixed part μ_{ij} which is usually a linear form in the predictors x_{ijk}, that is, $\mu_{ij} = \beta_1 x_{ij1} + \cdots + \beta_p x_{ijp}$, and an additive error ε_{ij}. In particular, when the process is started we have $\varepsilon_{i1} \sim N(0, \sigma^2)$. Now, at the second point in time the error $\varepsilon_{i2} = \rho \varepsilon_{i1} + \epsilon_{i2}$ depends on the error ε_{i1} through a given weight ρ and a new error component ϵ_{i2} which is independent of ε_{i1} and normally distributed, that is, $\epsilon_{i2} \sim N(0, \sigma_\epsilon^2)$. As the notation suggests, ρ is the correlation between ε_{i1} and ε_{i2}. This will be the result of the following arguments. Note that the variance of ϵ_{i2} is different from the variance of ε_{i1}. Now the variance of ε_{i2} is given as $\rho^2\sigma^2 + \sigma_\epsilon^2$. If one can assume constant variance of repeated observations, σ_ϵ^2 should be $(1-\rho^2)\sigma^2$. We assume therefore $\epsilon_{i2} \sim N(0, (1-\rho^2)\sigma^2)$. After having defined repeated observations in this way, we can calculate the correlation between ε_{i1} and ε_{i2},

$$
\begin{aligned}
Cov(\varepsilon_{i1}, \varepsilon_{i2}) &= E(\varepsilon_{i1}(\rho\varepsilon_{i1} + \varepsilon_{i2})) \\
&= E(\rho\varepsilon_{i1}^2) + E(\varepsilon_{i1}\epsilon_{i2}) \\
&= \rho\sigma^2.
\end{aligned}
$$

Now the correlation is obtained by simply dividing $Cov(\varepsilon_{i1}, \varepsilon_{i2})$ by the respective standard deviations (which are of course equal) and we get $Cor(\varepsilon_{i1}, \varepsilon_{i2}) = \rho$. We can see that the weight by which the error at time two is influenced by the error at time one is just the correlation between the two error terms, assuming, of course, constant variance of the observations.

We can generalize this idea to an arbitrary number of repeated observations. Let the error at time j be $\varepsilon_{ij} = \rho\varepsilon_{ij-1} + \epsilon_{ij}$ and the ϵ_{ij} independently $N(0, (1-\rho^2)\sigma^2)$ distributed. Notice that this distribution does not depend on time point j. The correlation between time j and time $j-k$ is then given by repeatedly using the arguments from above as $\rho^{j-k}, k = 1, \ldots, j-1$. In matrix notation the correlations between the observations of the ith individual can therefore be written as

$$
\mathbf{R_s} = \begin{pmatrix}
1 & \rho & \rho^2 & \cdots & \rho^n \\
\rho & 1 & \rho & \cdots & \rho^{n-1} \\
\vdots & \vdots & \vdots & \ddots & \vdots \\
\rho^n & \rho^{n-1} & \rho^{n-2} & \cdots & 1
\end{pmatrix}.
$$

It is interesting to note that in both models for building correlation into the within subject observations the whole correlation matrix is determined by only one parameter, ρ. There are many other models to explain this kind of correlation. For instance, one can build more sophisticated models by using more than one parameter, or one can build a mixture of both models explained above. Having parsimonious parameterized models, one can hope to get good parameter estimates. But this, of course, is only true if the correlational structure is correctly specified. In practice this structure is rarely known. If one is not willing to assume a certain model it is possible to estimate the whole covariance structure. It is then necessary to estimate all elements of the covariance matrix and not only one or possibly a few parameters according to the model for the covariance structure used. This technique is only practical when there are only a few repeated or correlated observations, because then the number of distinct elements in the covariance structure is small. If there are n repeated observations it is necessary to estimate $n(n+1)/2$ distinct entries in the matrix because the covariance matrix is symmetric. As this number increases quadratically with n this approach is only feasible when n is not too large. What "too large" exactly means is somewhat unclear as the amount of information in the data to estimate all $n(n+1)/2$ covariances depends on the number of replications as, by assumption, the covariance structure is the same for each replication. In brief, the estimation approach is useful when there are only few repeated observations relative to the total number of replications.

The nature of the problem can be further illustrated by considering a simple example from Dunlop (1994) where inferences based on ordinary least squares estimation can lead to incorrect conclusions if the correlation between observations is ignored. Suppose we have a sample of m individuals, half male and half female. Each individual is observed twice. y_{ij} represents the observation of individual i at time j, $j = 0, 1$. If we are interested in a possible group effect between males and females we could write down a linear model as

$$y_{ij} = \beta_0 + \beta_1 x_i + \epsilon_{ij},$$

where x_i is a dummy variable taking the value 0 for males and 1 for females. Interest lies in whether β_1 is significantly different from zero. $\epsilon_{ij} \sim N(\mu_i, \sigma^2 \mathbf{R})$, where $\mu_i = \beta_0 + \beta_1 x_i$, and the correlation matrix $\mathbf{R_s}$ for the ith individual has the simple form

$$\mathbf{R_s} = \begin{pmatrix} 1 & \rho \\ \rho & 1 \end{pmatrix}.$$

The covariance matrix $\mathbf{\Sigma_s}$ is therefore given by $\sigma^2 \mathbf{R_s}$. Note that the variance of the two observations is assumed to be constant. As before, $\mathbf{\Sigma}$

and $\mathbf{R}$ without subscripts refer to the covariance and correlation matrices of all observations, respectively, and have block diagonal form. Knowing the true $\mathbf{\Sigma}$, weighted least squares would result in the BLUE, that is, best linear unbiased estimate, of $\boldsymbol{\beta} = (\beta_0, \beta_1)$. Suppose for the moment the true covariance matrix $\mathbf{\Sigma}$ is known. Although the ordinary least squares estimate is not the best estimate we can possibly get, it is well known that it is unbiased regardless of the true correlation structure. This is not the case for the variance estimate of β_1. Because the OLS estimate is

$$\hat{\boldsymbol{\beta}} = (\mathbf{X}'\mathbf{X})^{-1}\mathbf{X}'\mathbf{y}$$

the variance can be calculated by linear means as

$$\begin{aligned} Var(\hat{\boldsymbol{\beta}}) &= (\mathbf{X}'\mathbf{X})^{-1}\mathbf{X}'Var(\mathbf{y})\mathbf{X}(\mathbf{X}'\mathbf{X})^{-1} \\ &= \sigma^2(\mathbf{X}'\mathbf{X})^{-1}\mathbf{X}'\mathbf{R}\mathbf{X}(\mathbf{X}'\mathbf{X})^{-1}, \end{aligned}$$

which would only be identical with the covariance matrix of $\hat{\boldsymbol{\beta}}$ using OLS if $\mathbf{R} = \mathbf{I}$, with $\mathbf{I}$ being the identity matrix. Exploiting the simple structure of $\mathbf{X}$, one obtains $var(\hat{\beta}_1)$ explicitly as $var(\hat{\beta}_1) = 2\sigma^2(1+\rho)/m$. Ignoring the correlation between the two repeated measures is equivalent to setting $\rho = 0$. The estimated variance would then incorrectly be estimated as $var(\hat{\beta}_1) = 2\sigma^2/m$. With a positive correlation ($\rho > 0$), which is typical for longitudinal data, the variance estimate of the interesting parameter β_1 will therefore be too small, resulting in progressive decisions concerning the group effect; that is, the null hypothesis of no group effect will be rejected too often. A similar example which models y_{ij} dependent on time leads to false variance estimates of the parameters as well, as was pointed out by Dunlop (1994).

This example further clears the point that the covariance structure of the observations has to be taken into account in order to perform a correct statistical analysis concerning inferences about the parameter vector $\boldsymbol{\beta}$.

14.2   Robust Modeling of Longitudinal Data

This approach was developed by Diggle, Liang, and Zeger (1994). Before describing this approach, let us first recall some facts about weighted least squares estimation. The WLS estimator, $\hat{\beta}$, is defined as the value for β which minimizes the quadratic form

$$(y - X\beta)'W(y - X\beta),$$

and the solution is given as

$$\hat{\beta} = (\mathbf{X}'\mathbf{W}\mathbf{X})^{-1}\mathbf{X}'\mathbf{W}\mathbf{y}.$$

The expected value of $\hat{\beta}$ is easily calculated using $E(\mathbf{y}) = X\beta$ to give

$$\begin{aligned} E(\hat{\beta}) &= E((\mathbf{X}'\mathbf{W}\mathbf{X})^{-1}\mathbf{X}'\mathbf{W}\mathbf{y}) \\ &= (\mathbf{X}'\mathbf{W}\mathbf{X})^{-1}\mathbf{X}'\mathbf{W}E(\mathbf{y}) \\ &= (\mathbf{X}'\mathbf{W}\mathbf{X})^{-1}\mathbf{X}'\mathbf{W}\mathbf{X}\beta \\ &= \beta; \end{aligned}$$

that is, the WLS estimator of $\hat{\beta}$ is unbiased for β whatever the choice for the weight matrix $\mathbf{W}$ will be. The variance of $\hat{\beta}$ is given as

$$var(\hat{\beta}) = (\mathbf{X}'\mathbf{W}\mathbf{X})^{-1}\mathbf{X}'\mathbf{W}\mathbf{\Sigma}\mathbf{W}\mathbf{X}(\mathbf{X}'\mathbf{W}\mathbf{X})^{-1},$$

where $\mathbf{\Sigma} = var(\mathbf{y})$ is the covariance matrix. The best linear unbiased estimator of β (which is also the maximum likelihood estimator), that is, the estimator with the smallest variance, is obtained by using $\mathbf{\Sigma}^{-1} = \mathbf{W}$. The formulas for calculating the estimator of β and its variance are then given by

$$\hat{\beta} = (\mathbf{X}'\mathbf{\Sigma}^{-1}\mathbf{X})^{-1}\mathbf{X}'\mathbf{\Sigma}^{-1}\mathbf{y}$$

and

$$Var(\hat{\beta}) = (\mathbf{X}'\mathbf{\Sigma}^{-1}\mathbf{X})^{-1}.$$

It is conceptually important for the following procedure to distinguish the weight matrix $\mathbf{W}$ from the covariance matrix $\mathbf{\Sigma}$. Because the true covariance matrix $\mathbf{\Sigma}$ is usually unknown in practice, the question arises whether using a suboptimal weight matrix would dramatically change the variance estimates of $\hat{\beta}$. This is often not the case. This can be seen using the above example from Dunlop (1994). Although Σ is usually unknown, for the following argument we keep Σ in the formula, replacing only the weight matrix $\mathbf{W}$. This is done to investigate the change in the variance for $\hat{\beta}$ if the optimal weight matrix is not used. We could interpret the Dunlop example as having used the identity matrix as the weight matrix in the formula that gives the variance of the OLS estimator. It could be asked as to how much the variance would change if we used the optimal weight matrix Σ^{-1} instead of the identity matrix. Interestingly, the variance for β_1 does not change at all. Therefore, we say that there is no loss in efficiency if we use a "false" weight matrix. This, unfortunately, is not always the case, especially when the used weight matrix totally misses the structure of the optimal weight matrix Σ^{-1}.

The main idea for the robust estimation approach is given in the following paragraph. By using a weight matrix which is not optimal but reasonably close to the true Σ^{-1} our inferences about the β vector are valid. In fact it can be shown that as long as we substitute a consistent estimate for the covariance matrix Σ the validity of our conclusions about β does not depend on the weight matrix $\mathbf{W}$. Choosing a suboptimal weight matrix affects only the efficiency of our inferences about β; that is, hypothesis tests and confidence intervals for the regression coefficient will be asymptotically correct. A consistent estimate of the true covariance matrix Σ means that as the number of observations increases, that is, the amount of information from the sample about the unknown parameters increases, the probability that the sample estimates of the elements of Σ are close to the true but unknown values approaches one. In brief, having a consistent estimate for Σ and a reasonable weight matrix $\mathbf{W}$, one is doing a WLS analysis and proceeds as if $\hat{\beta} \sim N(\beta, \hat{V}_W)$, where

$$\hat{V}_W = (\mathbf{X}'\mathbf{W}\mathbf{X})^{-1}\mathbf{X}'\mathbf{W}\hat{\Sigma}\mathbf{W}\mathbf{X}(\mathbf{X}'\mathbf{W}\mathbf{X})^{-1},$$

which is just the formula from above for $var(\hat{\beta})$ with Σ replaced by the consistent estimate $\hat{\Sigma}$. Subscript w indicates that the estimate is dependent on the specific choice of the weight matrix.

Now having come so far, the question that remains to be dealt with is how a consistent estimate for Σ can be obtained. Generally, this is referred to as variance component estimation. Two prominent methods for doing this are maximum likelihood (ML) and restricted maximum likelihood

(REML), estimation. The derivation of equations for these methods are beyond the scope of this book. A derivation of the equations that need to be solved for calculating the ML or the REML estimate of Σ can be found in Diggle et al. (1994). Normally these equations can only be solved numerically, but for the case of a designed experiment the computation simplifies considerably. Recall that in a designed experiment the values of the covariates are the same for each observation in a treatment group. This is typically not the case for observational data where there is no control over these values. Accounting for the various treatment groups, the notation is as follows. There are g treatment groups. In each group there are m_h individuals, each of which was repeatedly observed at n points in time. The complete set of measurements can be represented by y_{hij}, where $h = 1, \ldots, g; i = 1, \ldots, m_h; j = 1, \ldots, n;$ and $m = \sum_{i=1}^{g} m_i$.

The consistent REML estimate of Σ can then be obtained by first calculating the means over the observations within each of the g treatment group for each observation point, that is, $\bar{y}_{hj} = \frac{1}{m_h} \sum_{i=1}^{m_h} y_{hij}, j = 1, \ldots, n$. Let $\bar{\mathbf{y}}'_{\mathbf{h}} = (\bar{y}'_{h1}, \bar{y}'_{h2}, \ldots, \bar{y}'_{hn})$ be the mean response vector for group h and $\mathbf{y}_{\mathbf{h}i}$ the response vector for the ith individual in treatment group h. $\hat{\Sigma}_s$ is then obtained by calculating $(\mathbf{y}_{\mathbf{h}i} - \bar{\mathbf{y}}_{\mathbf{h}})(\mathbf{y}_{\mathbf{h}i} - \bar{\mathbf{y}}_{\mathbf{h}})'$ for each individual i. Note that this outer vector product results in a matrix. Summing all these matrices together and dividing the sum by $m - g$ gives $\hat{\Sigma}_s$. Formally this is

$$\hat{\Sigma}_s = \frac{1}{m - g} \sum_{h=1}^{g} \sum_{i=1}^{m_h} (\mathbf{y}_{\mathbf{h}i} - \bar{\mathbf{y}}_{\mathbf{h}})(\mathbf{y}_{\mathbf{h}i} - \bar{\mathbf{y}}_{\mathbf{h}})'.$$

$\hat{\Sigma}_s$ can be obtained, for instance, by matrix calculations using S-Plus or the matrix language of the standard statistical packages, for example, SAS-IML. This matrix can also be obtained by standard statistical packages as the error covariance matrix from a multivariate analysis of variance, treating the repeated observations from each individual as a multivariate response. To calculate $\hat{\Sigma}_s$ this way the observations y_{hij} have to be arranged as a matrix and not as a vector as described above. Note also that $(\mathbf{y}_{\mathbf{h}i} - \bar{\mathbf{y}}_{\mathbf{h}})$ is just the vector of residuals for the repeated observations of individual i from group h.

Having obtained the consistent estimate of Σ, a weight matrix $\mathbf{W}$ must be determined for weighted least squares analysis. Recall that Σ is just a block diagonal matrix with $\hat{\boldsymbol{\Sigma}}_s$ as the blocks. This is demonstrated by the following data example. We use a simulated data set. We are then able to evaluate the efficiency of our estimates obtained from the robust approach.

14.3   A Data Example

The data should be thought of as being obtained from 100 individuals at five different points in time. The sample is divided into two groups. The first 50 cases are females and the other 50 males. The time interval between repeated observations is the same. The correlation for the multivariate normal distribution of every individual follows a first-order Markov process with $\rho = 0.9$, that is, the error of observation separated by one, two, three and four time intervals is $0.9, 0.9^2 = 0.81, 0.9^3 = 0.729$, and $0.9^4 = 0.656$, respectively. The correlation is high but decreases with time. The mean response profiles which are added to the error are increasing functions of time with the difference between the groups getting smaller. Finally, the data are rounded to the nearest integer. The data set is presented in Appendix E.2 to allow the reader to reproduce the results.

Such data are fairly typical for social science data. An example would be that students fill out the same questionnaire in weekly intervals starting five weeks before an important examination to measure state anxiety. The research interest would then be (1) whether there are gender differences in state anxiety and (2) whether these differences change over time in different ways for the two groups; that is, whether there is an interaction between time and gender. First of all we can plot each individuals response profile (see Figure 14.1).

This plot is too busy to be useful. A useful plot presents the mean response profiles (see Figure 14.2). We see that state anxiety is steadily increasing and that the scores are higher for females than for males but that the difference decreases with time.

Now we first estimate $\hat{\boldsymbol{\Sigma}}$, because looking at this matrix could give us an idea what a reasonable choice for the weight matrix $\mathbf{W}$ would be. Remember that the optimal weight matrix is $\mathbf{W} = \boldsymbol{\Sigma}^{-1}$, so that if we look at the estimate of $\boldsymbol{\Sigma}$ we could possibly infer what a likely covariance

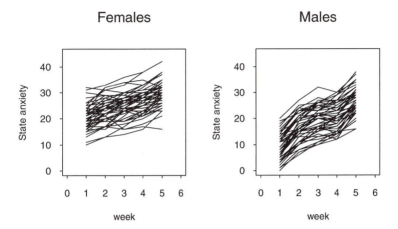

Figure 14.1: Development of state anxiety in females and males in the five weeks before an examination.

structure will be and then invert it to get $\mathbf{W}$. Recall also that one useful characteristic of this procedure is that only the efficiency of the inferences of $\boldsymbol{\beta}$ is decreased, but the validity of the inferences is not affected by a misspecification of $\mathbf{W}$. Perhaps the easiest way to calculate $\hat{\boldsymbol{\Sigma}}_s$ by matrix operations is to reorganize the $\mathbf{y}$ vector into a $\mathbf{Y}$ matrix with individuals as rows and repeated observations as columns. In the present example this results in a 100×5 matrix. We then have to build a design matrix $\mathbf{X}$ according to a linear model including the group effect. (If the groups were defined by some factorial design the design matrix should contain all main effects and interactions in the corresponding analysis of variance.) For this example the design matrix is

$$\mathbf{X} = \begin{pmatrix} 1 & 0 \\ 1 & 0 \\ 1 & 0 \\ \vdots & \vdots \\ 1 & 1 \\ 1 & 1 \\ 1 & 1 \end{pmatrix},$$

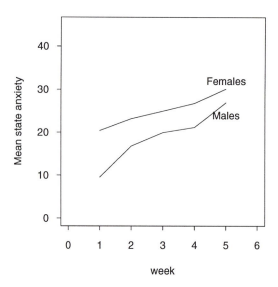

Figure 14.2: Average development of state anxiety in the two gender groups.

where the first column is the usual intercept and the second column dummy codes the two groups. It consists of 50 zeros for the 50 males and 50 ones for the 50 females.

The estimated covariance matrix is then given by the following matrix expression:

$$\hat{\boldsymbol{\Sigma}} = \frac{1}{m-g} \mathbf{Y}'(\mathbf{I} - \mathbf{X}(\mathbf{X}'\mathbf{X})^{-1}\mathbf{X}')\mathbf{Y}.$$

For these data one obtains (because the matrix is symmetric, only the diagonal and the lower diagonal elements will be given)

$$\hat{\boldsymbol{\Sigma}} = \begin{pmatrix} 25.191 & & & & \\ 22.515 & 25.606 & & & \\ 20.673 & 22.817 & 25.224 & & \\ 19.194 & 21.029 & 23.051 & 25.367 & \\ 16.127 & 17.668 & 20.072 & 22.608 & 25.850 \end{pmatrix}.$$

This reflects quite well the true covariance matrix, which is known to be

$$\hat{\Sigma} = \begin{pmatrix} 25.000 & & & & \\ 22.500 & 25.000 & & & \\ 20.250 & 22.500 & 25.000 & & \\ 18.225 & 20.250 & 22.500 & 25.000 & \\ 16.402 & 18.225 & 20.250 & 22.500 & 25.000 \end{pmatrix}$$

since the data were simulated. Hence the optimal weight matrix can be determined as the inverse of the true covariance matrix Σ. The form of W is quite interesting. It is well known that a first-order Markov process, according to which the data were simulated, has the property that given the present, the past and future are independent. This conditional independence relationship between observations, that are more than one time point apart, is reflected by the zero elements in the inverse of the covariance matrix; for details see Whittaker (1990).

The next step is to model the mean response profiles for the two groups. This can be done by polynomial regression. As can be seen from the plot, although there is some curvature in the response profile for the male group, a linear regression describes the mean responses over time quite well. If there are more time points, polynomial regression can be used. Now we set up the models for the males and females. They are

$$\begin{aligned} E(y_{Mj}) &= \beta_0 + \beta_1 t_j \\ E(y_{Fj}) &= \beta_0 + \beta_1 t_j + \tau + \gamma t_j. \end{aligned}$$

Of course, the model can be rewritten as

$$\begin{aligned} E(y_{Fj}) &= \beta_0 + \beta_1 t_j + \tau + \gamma t_j \\ &= (\beta_0 + \tau) + (\beta_1 + \gamma) t_j \\ &= \tilde{\beta}_0 + \tilde{\beta}_1 t_j, \end{aligned}$$

which shows that we are actually fitting two separate regression lines, one for each of the two groups. (The values of t_j were assumed for simplicity to have the values $1, \ldots, 5$.) The design matrix X for doing the regression has 500 rows and 5 columns. It has the following form, where the first 4 rows correspond to the five repeated observations of the first male. The second complete block of rows which are rows 251–255 in the complete matrix, correspond to the five repeated observations of the first female. These two blocks of five repeated observations are repeated 50 times for the male and 50 times for the female group, respectively, and one obtains

$$
\mathbf{X} = \begin{pmatrix}
1 & 1 & 1 & 1 \\
1 & 2 & 1 & 2 \\
1 & 3 & 1 & 3 \\
1 & 4 & 1 & 4 \\
1 & 5 & 1 & 5 \\
\vdots & \vdots & \vdots & \vdots \\
1 & 1 & 0 & 0 \\
1 & 2 & 0 & 0 \\
1 & 3 & 0 & 0 \\
1 & 4 & 0 & 0 \\
1 & 5 & 0 & 0 \\
\vdots & \vdots & \vdots & \vdots
\end{pmatrix}.
$$

$\mathbf{W}$ and $\hat{\boldsymbol{\Sigma}}$ are both block diagonal matrices and have 500 rows and 500 columns. If we use the optimal weights and the consistent estimate of $\boldsymbol{\Sigma}$ to fit the model and calculate standard errors by weighted least squares, the estimated regression coefficients (with standard errors in parentheses) are $\hat{\beta}_0 = 5.362(0.788), \hat{\beta}_1 = 4.340(0.153), \hat{\tau} = 12.635(1.114), and\, \hat{\gamma} = -1.928(0.217)$. The p values for the four hypotheses are often of interest. Tests for the hypotheses $H_0 : \beta_0 = 0; H_0 : \beta_1 = 0; H_0 : \tau = 0;$ and $H_0 : \gamma = 0$ can be derived from the assumption that $\hat{\boldsymbol{\beta}} \sim N(\boldsymbol{\beta}, \hat{\mathbf{V}}_{\boldsymbol{W}})$. In the example the $\boldsymbol{\beta}$ vector is given by $\boldsymbol{\beta}' = (\beta_0, \beta_1, \tau, \gamma)$. As the true covariance matrix is typically unknown, tests derived from the above assumption will only be asymptotically correct. If, for instance, $H_0 : \beta_0 = 0$ is true, then $\hat{\beta}_0/\hat{\sigma}(\hat{\beta}_0)$ will be approximately standard normally distributed. The same holds for the other three hypothesis tests. For hypothesis tests involving two or more parameters simultaneously see Diggle et al. (1994, p. 71). We calculate the observed z values, which should be compared to the standard normal distribution as $z(\beta_0) = 5.362/0.788 = 6.80, z(\beta_1) = 4.34/0.153 = 28.37, z(\tau) = 12.635/1.114 = 11.34,$ and $z(\gamma) = -1.928/0.217 = -8.88$.

All observed z values lie in the extreme tails of the standard normal distribution so that all calculated p values for one- or two-sided tests are virtually zero. The conclusions reached by this analysis are that, as can be seen from the plot, anxiety increases as the examination comes nearer ($\beta_1 > 0$) and the anxiety level for females is generally higher than that for males ($\tau > 0$). But there is also an interaction with time ($\gamma < 0$) showing that the difference between males and females decreases with time. The effect of τ should only be interpreted with great caution as one has to take the differential effect of time γ into account.

One benefit from using simulated data with known characteristics is that one can compare results with the OLS analysis of the data, that is, $\mathbf{W} = \mathbf{I}$, keeping the estimate of $\boldsymbol{\Sigma}$ in the formula to estimate the standard errors. This analysis yields the following values: $\beta_0 = 7.142(0.803), \beta_1 = 3.918(0.153), \tau = 11.089(1.135), and \gamma = -1.626(0.216)$. While the coefficients are slightly different, the obtained standard errors are nearly the same. Hypothesis tests follow as before with the result that all four parameter estimates are highly significant. If we look at the two plots in Figures 14.3 and 14.4 for the mean response profiles including the estimated regression lines for the optimal and the OLS analyses we can see that both analysis yield very similar results.

The conclusions of the analysis would surely not change. We see that the estimated standard errors are quite robust against a misspecification of $\mathbf{W}$. As the estimated covariance matrix is quite close to the true one, we would not expect the results to change for $\mathbf{W} = \boldsymbol{\Sigma}^{-1}$ as much as for $\mathbf{W} = \mathbf{I}$. But this cannot be recommended generally. Normally we use a guess for $\mathbf{W}$, which we hope captures the important characteristics. This is done by looking at the estimated covariance matrix and investigating how the covariances change with time. For the estimated covariance matrix it is seen that the ratio of two consecutive points in time is about 0.9. From this we could build a covariance matrix (which is, in this instance, the true correlation matrix) and invert it to obtain our guess for $\mathbf{W}$. Note that the results of the analysis do not change if we multiply $\mathbf{W}$ by an arbitrary constant. Thus it is only necessary to obtain a guess for the true covariance matrix up to an arbitrary constant for the guess of $\mathbf{W}$ as the inverse of the guess for $\boldsymbol{\Sigma}$.

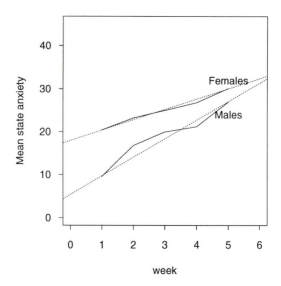

Figure 14.3: Mean response profiles and regression lines for both groups calculated using the optimal weight matrix.

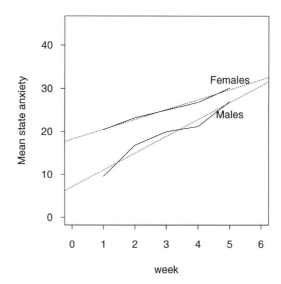

Figure 14.4: Mean response profiles and regression lines for both groups using the identity matrix as the weight matrix.

Chapter 15

PIECEWISE REGRESSION

Thus far, we have assumed that the regression line or regression hyper-plane is uniformly valid across the entire range of observed or admissible values. In many instances, however, this assumption may be hard to justify. For instance, one often reads in the newspapers that cigarettes damage organisms only if consumption goes beyond a certain minimum. A regression analysis testing this hypothesis would have to consider two slopes: one for the number of cigarettes smoked without damaging the organism, and one for the higher number of cigarettes capable of damaging the organism.

Figure 15.1 displays an example of a regression analysis with two slopes that meet at some cutoff point on X. The left–hand slope is horizontal; the right–hand slope is positive.

This chapter presents two cases of piecewise regression (Neter et al., 1996; Wilkinson, Blank, & Gruber, 1996). The first is piecewise regression where regression lines meet at the cutoff (*Continuous Piecewise Regression*; see Figure 15.1). The second is piecewise regression with a gap between the regression lines at the cutoff (*Discontinuous Piecewise Regression*).

These two cases share in common that the cutoff point is defined on the predictor, X. Cutoff points on Y would lead to two or more regression

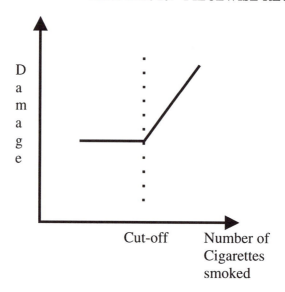

Figure 15.1: Continuous piecewise regression with one cutoff point.

lines that each cover the entire range of X values. The examples given in the following sections also share in common that there is only one cutoff point. The methods introduced for parameter estimation can be applied in an analogous fashion for estimation of parameters for problems with multiple cutoff points.

15.1   Continuous Piecewise Regression

We first consider the case where

1. piecewise regression is continuous, and

2. the cutoff point is known.

Let x_c be the cutoff point on X. Then, the model of *Simple Continuous Piecewise Regression* can be described as

$$Y = b_0 + b_1 X + b_2 (X - x_c) X_2 + \text{Residual}, \qquad (15.1)$$

where X is the regular predictor variable, and X_2 is a coding variable that assumes the following two values:

$$X_2 = \begin{cases} 1 & \text{if } X > x_c, \\ 0 & \text{otherwise.} \end{cases}$$

The effect of the second coding variable is that when b_2 is estimated, only values greater than x_c are considered.

The design matrix for this type of piecewise regression contains three vectors, specifically, the constant vector, the vector of predictor values, x_i, and a vector that results from multiplying X_2 with the difference $(X - x_c)$.

The following design matrix presents an example. The matrix contains data for six cases, the second, the third, and the fourth of which have values greater than x_c. For each case, there is a constant, a value for X, and the value that results from $X_2(X - x_c)$.

$$\mathbf{X} = \begin{pmatrix} 1 & x_{1,11} & 0 \\ 1 & x_{1,21} & (x_{1,21} - x_c)x_{2,21} \\ 1 & x_{1,31} & (x_{1,31} - x_c)x_{2,31} \\ 1 & x_{1,41} & 0 \\ 1 & x_{1,51} & (x_{1,51} - x_c)x_{2,51} \\ 1 & x_{1,41} & 0 \end{pmatrix}. \tag{15.2}$$

The zeros in the last column of X result from multiplying $(x_{1,ij} - x_c)$ with $x_{2,ij} = 0$, where i indexes subjects and j indexes variables.

The following data example analyzes data from the von Eye et al. (1996) study. Specifically, we regress the cognitive complexity variables, Breadth $(CC1)$, and Overlap (OVC), for $n = 29$ young adults of the experimental group. Figure
reffi:piecew2 displays the Breadth by Overlap scatterplot and the OLS regression line. The regression function estimated for these data is

$$OVC = 0.90 - 0.03 * CC1 + \text{Residual.}$$

This equation explains $R^2 = 0.60$ of the criterion variance and has a significant slope parameter $(t = -6.40, p < 0.01)$. Yet, the figure shows that there is no sector of the predictor, $CC1$, where this regression line provides a particularly good approximation of the data.

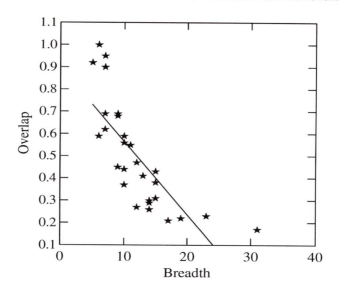

Figure 15.2: Regression of Overlap of Cognitive Complexity, OVC, on Breadth of Cognitive Complexity, CC1.

Therefore, we test the hypothesis that the steep slope of the regression of *OVC* on *CC1* that characterizes the relationship for values $CC1 \leq 15$ is followed by a much flatter slope for values $CC1 > 15$. To test this hypothesis we perform the following steps:

1. Create the variable X_2 with values as follows:

$$X_2 = \begin{cases} 1 & \text{if CC1} \leq 15, \\ 0 & \text{otherwise.} \end{cases}$$

2. Create the product $(CC1 - 15) * X_2$.

3. Estimate parameters for the regression equation

$$\text{OVC} = b_0 + b_1 * \text{CC1} + b_2 * (\text{CC1} - 15) * X_2 + \text{Residual.}$$

The following parameter estimates result:

$$\text{OVC} = 1.130 + 0.056 * \text{CC1} + 0.002 * (\text{CC1} - 15) * X_2 + \text{Residual.}$$

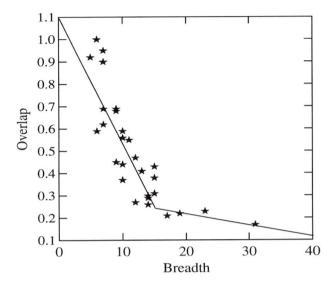

Figure 15.3: Piecewise regression of Overlap on Breadth with known cutoff point.

All of the parameters are significant, with the following three t values for b_0, b_1, and b_2, respectively: 14.98, 4.35, and -8.40. A portion of $R^2 = 0.769$ of the criterion variance is explained.

Figure 15.3 displays the $CC1 * OVC$ scatterplot with the piecewise regression line.

There can be no doubt that the data are much better represented by the piecewise regression than by the straight line regression in Figure 15.2.

15.2   Discontinuous Piecewise Regression

In this section we

1. illustrate discontinuous piecewise regression and

2. show how curvilinear regression can be applied in piecewise regression.

While we show this using one example, curvilinear and discontinuous piecewise regression do not necessarily go hand in hand. There are many examples of curvilinear continuous piecewise regression (Wilkinson, Hill, Welna, & Birkenbeuel, 1992, Ch. 9.6), and there are approaches to linear piecewise regression that are discontinuous (Neter et al., 1996, Chap. 11).

Formula (15.1) presents a regression equation with two linear, additive components that can differ in slope. Specifically, these components are $b_1 X$ and $b_2(X - x_c)X_2$. This section illustrates how one (or both) of these linear components can be replaced by a curvilinear component. The regression model used in this section has the form

$$Y = b_0 + \sum_j b_j \xi_j + \text{Residual}, \qquad (15.3)$$

where ξ_j are regression functions that can be linear or curvilinear. Formula (15.1) is a special case of (15.3) in that the sum goes over two linear regression functions.

The following example uses the data from Figure 15.3 again. The figure suggests that there are segments of the predictor, *Breadth*, that are not particularly well covered. For example, there are four cases with values of Breadth < 7 and Overlap > 0.8 that are poorly represented by the left-hand part of the regression line. Because changing parameters of the linear, continuous piecewise regression line would lead to poor representations in other segments of the predictor, we select nonlinear regression lines. In addition, we allow these lines to be discontinuous, that is, be unconnected at the cutoff.

To optimize the fit for the regression of Overlap on Breadth of Cognitive Complexity we chose the function

$$\text{Overlap} = b_0 + b_1(\exp b_2(-\text{Breadth})) + b_3(\text{Breadth}^2) + \text{Residual}, \qquad (15.4)$$

where the cutoff between the two components of the piecewise regression is, as before, at Breadth $= 15$. The first component of the curvilinear piecewise regression is an exponential function of Breadth. The second component is a quadratic function of Breadth.

Technically, the function given in (15.4) can be estimated using computer programs that perform nonlinear OLS regression, for instance, SPSS, SAS, BMDP, and SYSTAT. The following steps are required:

1. Create indicator variable, X_2, that discriminates between predictor values before and after cutoff

2. Multiply indicator variable, X_2, with Breadth squared (see (15.4));

3. Estimate parameters for (15.4).

In many instances, even standard computer programs for OLS regression can be used, when all terms can be transformed such that a linear regression model can be estimated where all exponents of parameters equal 1 and parameters do not appear in exponents.

For the present example we estimate the parameters

$$\text{Overlap} = 0.29 + 2.22(\exp 0.23(-\text{Breadth}))$$
$$+ 0.00015(\text{Breadth}^2) + \text{Residual},$$

where the second part of the regression model, that is, the part with Breadth^2, applies to all cases with values of Breadth ≥ 15. The scatterplot of Breath * Overlap of Cognitive Complexity appears in Figure 15.4, along with the curvilinear, discontinuous piecewise regression line.

The portion of variance accounted for by this model is $R^2 = 0.918$, clearly higher than the $R^2 = 0.769$ explained by the model that only used linear regression lines. The figure shows that the two regression lines do not meet at the cutoff. When linear regression lines are pieced together, as was illustrated in the first data example (see Figure 15.3), one needs a separate parameter that determines the magnitude of the leap from one regression line to the other. In the present example with curvilinear lines, the curves do not meet at the cutoff and, thus, create a leap.

Extensions. Piecewise regression can be extended in various ways, two of which will be mentioned in this section. One first and obvious extension concerns the number of cutoffs. Consider a researcher that investigates the Weber–Fechner Law. This law proposes that differences in such physical characteristics of objects as weight or brightness can be perceived only if they are greater than a minimum proportion. This proportion is assumed to be constant:

$$\frac{\Delta R}{R} = \text{constant}.$$

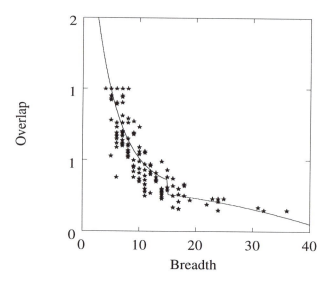

Figure 15.4: Curvilinear, discontinuous piecewise regression of Overlap on Breadth of Cognitive Complexity.

It is well known that this law is valid only in the middle of scales, not at the extremes. Therefore, a researcher investigating this law may consider two cut-offs rather than one. The first cutoff separates the lower extreme end of the scale from the rest. The second cutoff separates the upper extreme end from the rest.

The second extension concerns the use of repeated observations. Consider a researcher interested in the stability of effects of psychotherapy. This researcher may wish to model the observations made during the therapy using a first type of function, for example, a negatively decelerated curve as in the left-hand side of Figure 15.4. For the time after completion of therapy, this researcher may wish to use some other model.

Caveat. It is tempting and fun to improve fit using more and more refined piecewise regression models. There are not too many data sets that a data analyst that masters the art of nonlinear and piecewise estimation will not be able to depict very well. "Nonlinear estimation is an art... is rococo" (Wilkinson et al., 1992, p. 428). However, the artist needs guidance from theory. If there is no such guidance, models may provide impressive fit. However, that fit often reflects sample specificities rather than results that can be replicated or generalized.

Chapter 16

DICHOTOMOUS CRITERION VARIABLES

This chapter presents a simple solution for the case where both the predictor and the criterion are dichotomous, that is, only have two categories. Throughout this chapter, we assume that the predictor can be scaled at the nominal level. Alternatively, the predictor can be scaled at any higher level, but was categorized to reduce the number of levels.

As far as the criterion is concerned, scaling determines the sign and its interpretation of the slope parameter. If both the predictor variable and the criterion variable are scaled at least to the ordinal level, a sign can be meaningfully interpreted. For nominal level predictors or criteria, signs of regression slopes are arbitrary. It should be noted that, in many contexts, the distinction between scale levels does not have any consequences when variables are dichotomous. In the present context, scaling does make a difference.

More specifically, when the criterion variable is scaled at the ordinal level (or higher), the sign can be interpreted as follows

1. A positive sign suggests that the "low" predictor category allows one to predict the "low" criterion category and the "high" predictor

Table 16.1: *Scheme of a 2 x 2 Table*

Predictor	Criterion Values	
Categories	1	2
1	a	b
2	c	d

category allows one to predict the "high" criterion category.

2. A negative sign suggests that the "low – high" and the "high – low" categories go hand-in-hand.

The theoretical background for the method presented here is the well-known result that the ϕ coefficient of association between categorical variables can be shown to be a special case of Pearson's correlation coefficient, r. As is also well known, the relationship between r and the slope coefficient, b_1, is

$$b_1 = r \frac{s_x}{s_y},$$

where s_x is the standard deviation of X, and s_y is the standard deviation of Y. Thus, one can ask what the regression of one categorical onto another is. In this chapter we focus on dichotomous variables. Consider Table 16.1.

The cells of this 2 x 2 table, denoted by a, b, c, and d, contain the numbers of cases that display a certain pattern. For instance, cell a contains all those cases that display predictor category 1 and criterion value 1. From this table, the regression slope coefficients can be estimated as follows:

$$b_1 = \frac{ad - bc}{(a + b)(c + d)}. \tag{16.1}$$

The intercept can be estimated using (16.1) as follows:

$$b_0 = \frac{a + c}{n} - b_1 \frac{a + b}{n}. \tag{16.2}$$

Table 16.2: *Handedness and Victory Pattern in Tennis Players*

Handedness	Victories	
	1	2
r	41	38
l	4	17

In the following numerical example we predict number of victories in professional tennis, V, from handedness, H. We assign the following values: $V = 1$ for a below average number of victories and $V = 2$ for an above average number of victories, and $H = l$ for left-handers and $H = r$ for right-handers. A total of $n = 100$ tennis players is involved in the study. The frequency table appears in Table 16.2.

Inserting into (16.1) yields

$$b_1 = \frac{41 * 17 - 4 * 38}{(41 + 38)(4 + 17)} = 0.329.$$

Inserting into (16.2) yields

$$b_0 = \frac{41 + 4}{100} - 0.329\frac{41 + 38}{100} = 0.45 - 0.26 = 0.19.$$

Taken together, we obtain the following regression equation:

Victories $= 0.19 + 0.329 *$ Handedness $+$ Residual.

The interpretation of regression parameters from 2 x 2 tables differs in some respects from the interpretation of standard regression parameters. We illustrate this by inserting into the last equation. For $H = r = 1$ we obtain

$$y = 0.19 + 0.329 * 1 = 0.458,$$

and for $H = l = 0$ (notice that one predictor category must be given the numerical value 0, and the other must be given the value 1; this is a variant of dummy coding) we obtain

$$y = 0.19 + 0.329 * 0 = 0.19.$$

Obviously, these values are not frequencies. Nor are they values that the criterion can assume. In general, interpretation of regression parameters and estimated values in 2 x 2 regression is as follows:

1. The regression parameter β_0 reflects the conditional probability that the first criterion category is observed, given the predictor category with the value 0 or, in more technical terms,

$$\beta_0 = prob(Y = 1 | X = 0).$$

2. The regression slope parameter β_1 suggests how much the probability that $Y = 1$, given $X = 1$, differs from the probability that $Y = 1$, given $X = 0$.

Using these concepts we can interpret the results of our data example as follows:

- The intercept parameter estimate, $b_0 = 0.19$, suggests that the probability that left-handers score a below average numbers of victories is $p = 0.19$.

- The slope parameter estimate, $b_1 = 0.329$, indicates that the probability that right-handers score a below average numbers of victories is $p = 0.329$ higher than the probability for left-handers.

Chapter 17

COMPUTATIONAL ISSUES

This chapter illustrates application of regression analysis using statistics software packages for PCs. The package used for illustration is SYSTAT for Windows, Releases 5.02 and 7.0. Rather than giving sample runs for each case of regression analysis covered in this volume, this chapter exemplifies a number of typical cases. One sample set of data is used throughout this chapter. The file contains six cases and three variables. This chapter first covers data input. Later sections give illustrations of regression runs.

17.1 Creating a SYSTAT System File

SYSTAT can be used to input data. However, the more typical case is that the data analyst receives data on a file in, for instance, ASCII code. In this chapter we illustrate how to create a SYSTAT system file using an existing ASCII code data file. The file used appears below:

```
1   3   1
2   5   3
3   9   2
4  11   4
```

5   9   3
6 11   5

One of the main characteristics of this file is that data are separated by spaces. This way, SYSTAT can read the data without format statements. The first of the columns contains the values for Predictor1, the second column contains the values for Criterion, and the third column contains the values for Predictor2. For the following purposes we assume that the raw data are in the file "Raw.dat." SYSTAT expects an ASCII raw data file to have the suffix "dat."

To create a SYSTAT system file from this raw data ASCII file, one issues the following commands:

Command	Effect
Data	Invokes SYSTAT's Data module
Save Compdat	Tells SYSTAT to save data in file; "Compdat.sys," where the suffix "sys" indicates that this is a system file (recent versions of SYSTAT use the suffix "syd."
Input Predictor1, Criterion, Predictor2	Conveys variable names to program
Get Raw	Reads raw data from file "Raw.dat"
Run	Saves raw data in system file "Compdat.sys"; the system file also contains variable names

After these operations there exists a SYSTAT system file. This file is located in the directory that the user has specified for Save commands. For the following illustrations we assume that all files that we create and use are located in the default directory, that is C:\SYSTATW5\.

Before analyzing these data we create a graphical representation of the data, and we also calculate a correlation matrix. Both allow us to get a first image of the data and their characteristics. Figure 17.1 displays the response surface of the data in file "Compdat.sys."

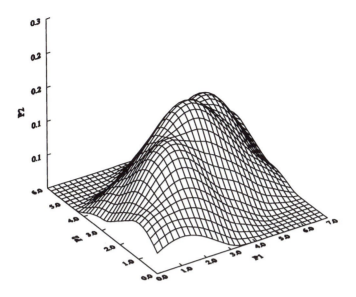

Figure 17.1: Response surface for sample data.

This response surface can be created by issuing the following commands:

Command	Effect
Use Compdat	Reads file "Compdat.sys"
click Graph, Options, Global	Provides print options for graphics
click Thickness = 3, Decimals to Print = 1, and select from the font pull down the British font	Specifies output characteristics
OK	Concludes selection of output characteristics
click Graph, 3-D, 3-D	Opens the menu for 3-D graphs
	continued on next page

assign Predictor1 to X, Predictor2 to Y, and Criterion to Z	Assigns variables to axes in 3-D representation
click Axes and insert in Label Fields for X, Predictor1; for Y, Predictor2; and for Z, Criterion; OK	Labels axes
click Surface and select the Kernel Smoother, OK	Specifies method for creating surface
OK	Starts creation of graph on screen
click File in Graph Window and select Print	Initiates printing process
increase Enlarge to 150%; OK	increases print output to 150% (results in output 6 inches in height; width depends on length of title)
select from the following window OK	Starts printing

Table 17.1 gives the variable intercorrelations and the Bonferroni-significance tests for these intercorrelations.

Table 17.1: *Matrix of Intercorrelations (Upper Triangle) and Bonferroni-Significance Values (Lower Triangle) for Predictor1, Criterion, and Predictor2*

	Predictor1	Criterion	Predictor2
Predictor1		0.878	0.832
Criterion	0.064		0.775
Predictor2	0.121	0.211	

The correlation table can be created by issuing the following commands:

Command	Effect
Use Compdat	Reads file "Compdat.sys"
click Stats, Corr, Pearson	Selects method for statistical analysis
click Bonferroni	Specifies that Bonferroni significance tests be performed; variable selection is not necessary if all variables are involved in correlation matrix
OK	Initiates computation and sends results to screen
highlight desired print output	
click File, Print, OK	Sends highlighted selection to printer

Both the response surface and the correlation table suggest that predictors and criteria are strongly intercorrelated. After Bonferroni adjustment, the correlations fail to reach significance. However, the portions of variance shared in common are high for all three variable pairs.

The following chapters illustrate application of regression analysis using SYSTAT when regressing the variable Criterion onto Predictor1, Predictor2, or both.

17.2 Simple Regression

This section illustrates application of simple linear regression. We regress Criterion onto Predictor1. To do this, we issue the following commands:

Command	Effect
Use Compdat	Reads file "Compdat.sys"
click Stats, MGLH	Initiates statistical analysis; selects the MGLH (Multivariate General Linear Hypotheses) module
click Regression	Selects regression module
assign Predictor1 to Independent; assign Criterion to Dependent	Specifies predictors and criteria in regression
OK	Starts calculations; sends results to screen

To print results we proceed as before. The following output results for the present example:

```
USE 'A:\COMPDAT.SYS'
SYSTAT Rectangular file A:\COMPDAT.SYS,
created Mon Nov 17, 1997 at 19:26:32, contains variables:
 P1            C              P2

>EDIT
>REGRESS
>MODEL C = CONSTANT+P1
>ESTIMATE

Dep Var: C
N: 6
Multiple R: 0.878
Squared multiple R: 0.771

Adjusted squared multiple R: 0.714
Standard error of estimate: 1.757

Effect    Coefficient   Std Error   Tol.    t     P(2 Tail)

CONSTANT    2.600        1.635        .     1.590    0.187
P1          1.543        0.420      1.000   3.674    0.021
```

Reading from the top, the output can be interpreted as follows:

- The top four lines tell us what file was opened, give the date and time, and name the variables in the file.

- The result display begins with (1) the dependent variable, (C)riterion, and gives (2) the sample size, 6, (3) the multiple R, and (4) the multiple R^2. When there is only one predictor, as in the present example, the multiple R and the Pearson correlation, r, are identical (see Table 17.1);

- The 13th line gives the adjusted multiple R^2, which is calculated as

$$R_a^2 = \frac{(m-1)(1-R^2)}{n-m},$$

where m is the number of predictors, and N is the sample size. R_a^2 describes the portion of variance accounted for in a new sample from the same population. At the end of the 13th line we find the standard error of the estimate, defined as the square root of the residual mean square.

- After a table header we find the parameters and their standard errors, the standardized coefficients ("betas" in SPSS); the tolerance value (see the section on multiple regression); the t value; and the two-sided tail probability for the t's;

To depict the relationship between Predictor1 and Criterion we create a graph with the following commands:

Command	Effect
Use Compdat	Reads file "Compdat.sys"
click Graph, Options, Global	Provides print options for graphics
click Thickness = 3, Decimals to Print = 1, and select from the font pull down the British font	Specifies output characteristics
	continued on next page

OK	Concludes selection of output characteristics
click Graph, Plot, Plot	Opens the menu for 2-D plots
assign Predictor1 to X and Criterion to Y	Assigns variables to axes
click Axes and insert in Label Fields: for X: Predictor1 and for Y: Criterion, OK	Labels axes
click Smooth and OK	Specifies type of regression line to be inserted in the graph; in the present example there is no need to make a decision, because a linear regression line is the default
click Symbol, select symbol to represent data points (we take the star), and select size for data points (we take size 2)	Specifies print options
OK	Starts creation of graph on screen
click File in Graph Window and select Print	Initiates printing process
increase Enlarge to 150%; OK	Increases print output to 150% (results in output that is 6 inches in height; width depends on length of title)

The resulting graph appears in Figure 17.2. Readers are invited to create the same regression output and the same figure using Criterion and Predictor2.

17.3   Curvilinear Regression

For the following runs suppose the researchers analyzing the data in file "Compdat.sys" derive from theory that a curvilinear regression slope may validly describe the the slope of Criterion. Specifically, the researchers

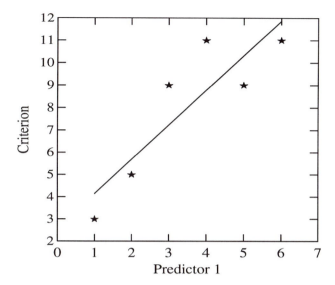

Figure 17.2: Simple linear regression of Criterion on Predictor1 of sample data.

assume that there may be a quadratic slope. To test this assumption the researchers perform the following two steps:

1. Create a variable that reflects a quadratic slope for six observation points.

2. Predict Criterion from this variable.

Let us call the variable that gives the quadratic slope Quad. Then, one performs the following steps in SYSTAT to test the assumption of a quadratic fit.

Step 1 is to create a coding variable for a quadratic slope:

Command	Effect
Use Compdat	Reads data file "Compdat.sys"
click Window, Worksheet	Opens a worksheet window that displays the data
	continued on next page

move cursor in top line, that is, the line that contains variable names, and move to the right, one cell past the last existing name	Opens array (column) for new entries
type "Quad"	Creates variable Quad
move cursor one step down, type "5," move cursor one step down, type "-1"	Insert values for variable Quad (values are taken from Kirk (1995, Table E10))
repeat until last value for variable Quad is inserted	Insert all polynomial coefficients; the last four coefficients are -4, -4, -1, 5
click File, select Save	Saves new file under same name, that is, under "Compdat.sys"
click File, select Close	Closes worksheet; reads new data; readies data for analysis

We now have a new data file. It contains the original variables, Predictor1, Predictor2, and Criterion, and the new variable, Quad, which carries the coefficients of a quadratic polynomial for six observation points. In Step 2, we predict Criterion from Quad.

Step 2 consists of testing the assumption that Criterion has a quadratic slope:

Command	Effect
Use Compdat	Reads data file "Compdat.sys"
click Stats, MGLH, and Regression	Specifies that OLS regression will be employed for data analysis
assign Criterion to Y and Quad to X	Defines predictor and criterion
click OK	Starts analysis and send results to screen

These steps yield the results displayed in the following output:

```
MODEL C = CONSTANT+QUAD
>ESTIMATE
```

```
Dep Var: C
N: 6
Multiple R: 0.356
Squared multiple R: 0.127
```

```
Adjusted squared multiple R: 0.0
Standard error of estimate: 3.433
```

Effect	Coefficient	Std Error	t	P(2 Tail)
CONSTANT	8.000	1.402	5.708	0.005
QUAD	-0.286	0.375	-0.763	0.488

Reading from the top of the regression table, we find the same structure of output as before. Substantively, we notice that the quadratic variable gives a poorer representation of the data than Predictor1. The variable explains no more than $R^2 = 0.127$ of the criterion variance. Predictor1 was able to explain $R^2 = 0.771$ of the criterion variance.

Asking why this is the case, we create a graph that displays the observed values and the values estimated by the regression model. To create this graph, we need the estimates created by the regression model with the quadratic predictor. The following commands yield a file with these estimates.

Command	Effect
Use Compdat	Reads data file "Compdat.sys"
click Stats, MGLH, and Regression	Selects regression analysis
assign Criterion to Y and Quad to X	Specifies regression predictor and criterion
select Save Residuals	Initiates saving of residuals in new file
	continued on next page

OK	Asks for file name and specification of type of information to save
type file name for residuals (we type "Quadres"), click OK	Specifies file name, readies module for calculations; there is no need for further selections because the default choice saves the information we need
OK	Starts calculations, sends residuals and diagnostic values to file "Quadres.sys"; sends regression results to screen
type "Data"	Invokes SYSTAT's Data module
type "Use Compdat Quadres"	Reads files "Compdat.sys" and "Quadres.sys" simultaneously; makes all variables in both files simultaneously available
type "Save Quadres2"	Creates file "Quadres2.sys" in which we save information from both files
type "Run"	Merges files "Compdat.sys" and "Quadres.sys" and saves all information in file "Quadres2.sys"

After issuing these commands, we have a file that contains the original variables, Predictor1, Predictor2, and Criterion; the variable Quad; and the estimates, residuals, and various residual diagnostic values (see Chapter 6). This is the information we need to create the graph that plots observed versus estimated criterion values. The following commands yield this graph:

Command	Effect
Use Quadres2	Reads file "Quadres2.sys"
click Graph, Options, Global	Provides print options for graphics
	continued on next page

click Thickness = 3, Decimals to Print = 1, and select from the font pull down the British font	Specifies output characteristics
OK	Concludes selection of output characteristics
click Graph, Plot, Plot	Opens the menu for 2D plots
assign Predictor1 to X, Criterion to Y, and Estimate also to Y	Assigns variables to axes
click Axes and insert in Label Fields; for X, Predictor1, and for Y, Criterion; OK	Labels axes
click Symbol, select symbol to represent data points (we take the star), and select size for data points (we take size 2)	Specifies print options
OK	Starts creation of graph on screen
click File in Graph Window and select Print	Initiates printing process
increase Enlarge to 150%; OK	Increases print output to 150% (results in output that is 6 inches in height; width depends on length of title)

The resulting graph appears in Figure 17.3. The figure indicates why the fit provided by the quadratic polynomial is so poor. When comparing the stars (observed data points) and the circles (expected data points) above the markers on the Predictor1 axis, we find that, with the exceptions of the third and the fifth data points, not one is well represented by the quadratic curve. Major discrepancies are apparent for data points 1, 2, and 7. What one would need for an improved fit is a quadratic curve tilted toward the upper right corner of the graph. The next section, on multiple regression, addresses this issue.

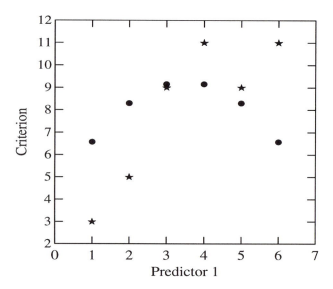

Figure 17.3: Observed data points (stars) and expected data points (circles) in quadratic regression.

17.4   Multiple Regression

The curvilinear analysis performed in Section 17.3 suggested a poor fit. The main reason identified for this result was that the quadratic curve needed a positive linear trend that would lift its right-hand side. To create a regression line that has this characteristic we now fit a complete quadratic orthogonal polynomial, that is, a polynomial that involves both a linear term and the quadratic term used in Section 17.3.

To create a coding variable that carries the coefficients of the linear orthogonal polynomial we proceed as in Section 17.3, where we created a coding variable that carried the coefficients for the quadratic polynomial. Readers are invited to do this with their own computers. Let the name of the new coding variable be Lin. It has values -5, -3, -1, 1, 3, and 5. We store the data set, now enriched by one variable, in file "Quadres2.sys."

Using the linear and the quadratic orthogonal polynomials we can (1) calculate a multiple regression and (2) create a graph that allows us to compare the results from the last chapter with the results from this chapter. The following commands yield a multiple regression:

Command	Effect
Use Quadres2	Reads file "Quadres2.sys"
click Stats, MGLH	Initiates statistical analysis; selects the MGLH (Multivariate General Linear Hypotheses) module
click Regression	Selects Regression module
assign Lin and Quad to Independent; assign Criterion to Dependent	Specifies predictors and criteria in regression
OK	Starts calculations; sends results to screen

The resulting printout appears below.

```
MODEL C = CONSTANT+LIN+QUAD

>ESTIMATE

Dep Var: C
N: 6
Multiple R: 0.948
Squared multiple R: 0.898

Adjusted squared multiple R: 0.831
Standard error of estimate: 1.352
```

Effect	Coefficient	Std Error	Tol.	t	P(2 Tail)
CONSTANT	8.000	0.552	.	14.491	0.001
LIN	0.771	0.162	1.00	4.773	0.017
QUAD	-0.286	0.148	1.00	-1.936	0.148

```
                 Analysis of Variance

Source        SS      df    MS       F        P

Regression  48.514    2   24.257   13.266   0.032
Residual     5.486    3    1.829
```

The output for multiple regression follows the same scheme as for simple regression. Yet, there are two characteristics of the results that are worth mentioning. The first is the Tolerance value. Tolerance is an indicator of the magnitude of predictor intercorrelations (see the VIF measure in Section 8.1). It is defined as

$$\text{Tolerance}_j = 1 - R_i^2, \quad \text{over all } i \neq j,$$

or, in words, the Tolerance of Predictor i is the multiple correlation between this predictor and all other predictors. It should be noticed that the criterion is not part of these calculations. Tolerance values can be calculated for each predictor in the equation. When predictors are completely independent, as in an orthogonal design or when using orthogonal polynomials, the Tolerance assumes a value of Tolerance = 1 - 0 = 1. This is the case in the present example. Tolerance values decrease with predictor intercorrelation.

The second characteristic that is new in this printout is that the ANOVA table is no longer redundant. ANOVA results indicate whether the regression model, as a whole, makes a significant contribution to explaining the criterion. This can be the case even when none of the single predictors make a significant contribution. The F ratio shown is no longer the square of any of the t values in the regression table.

Substantively, we note that the regression model with the complete quadratic polynomial explains $R^2 = 0.898$ of the criterion variance. This is much more than the quadratic polynomial alone was able to explain. The linear polynomial makes a significant contribution, and so does the entire regression model.

To compare results from the last chapter with these results we want to draw a figure that displays two curves, that from using only the quadratic term and that from using the complete quadratic polynomial. We issue the following commands:

Command	Effect
Use Quadres2	Reads file "Quadres2.sys"
click Graph, Options, Global	Provides print options for graphics
click Thickness = 3, Decimals to Print = 1, and select from the font pull down the British font (or whatever font pleases you)	Specifies output characteristics
OK	Concludes selection of output characteristics
click Graph, Plot, Plot	Opens the menu for 2D plots
assign Estimate and Predictor1 to X, and Criterion to Y	Assigns variables to axes
click Axes and insert in Label Fields, for X, Predictor1, and for Y, Criterion; OK	Labels axes
click Symbol, select symbol to represent data points (we take the diamond and the star, that is, numbers 1 and 18), and select size for data points (we take size 2)	Specifies print options
click Smooth, select Quadratic	Specifies type of regression line
OK, OK	Starts drawing on screen
click File in Graph Window and select Print	Initiates printing process
increase Enlarge to 150%; OK	Increases size of print output and starts the printing

Figure 17.4 displays the resulting graph.

The steeper curve in the figure represents the regression line for the

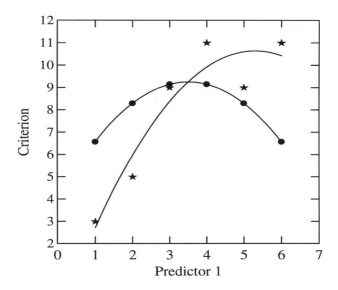

Figure 17.4: Comparing two quadratic regression lines.

complete quadratic polynomial. The curve that connects the circles represents the regression line from the last section. Quite obviously, the new regression line gives a much closer representation of the original data points than the earlier regression line. Adding the linear component to the polynomial has provided a substantial increase in fit over the quadratic component alone.

17.5   Regression Interaction

This section illustrates how to estimate regression interaction models. We use file "Compdat.sys" with the variables Predictor1, Predictor2, and Criterion. The sample model analyzed here proposes that the slope of the line that regresses Criterion on Predictor1 depends on Predictor2. More specifically, we investigate the model

$$\text{Criterion} = \text{Constant} + \text{Predictor1} + \text{Predictor1} * \text{Predictor2}$$

(see Chapter 10).

In order to be able to include the multiplicative term in a model that can be analyzed using SYSTAT's Regression module, one must create a new variable. This variable results from element-wise multiplying Predictor1 with Predictor2. However, there is a simpler option. It involves using the more general MGLH General Linear Model module. This model allows one to have the program create the multiplicative variable. The following commands need to be issued:

Command	Effect
Use Compdat	Reads data file "Compdat.sys"
click Stats, MGLH, and General linear model	Selects GLM module
assign Criterion to Dependent and Predictor1 to Independent	Specifies criterion and predictor of regression
highlight both Predictor1 and Predictor2 and click Cross under the Independent variable box	The multiplicative term Predictor1 * Predictor2 will appear as new independent variable, thus completing the specification of the regression model
click OK	Starts data analysis; sends output to screen

The following output results from these commands:

```
MODEL C = CONSTANT+P1+P2+p1*p2

>ESTIMATE

Dep Var: C
N: 6
Multiple R: 0.904
Squared multiple R: 0.817

Adjusted squared multiple R: 0.542
Standard error of estimate: 2.224
```

Effect	Coefficient	Std Error	Tol.	t	P(2 Tail)
CONSTANT	0.048	4.190	.	0.011	0.992
P1	2.125	1.544	0.119	1.376	0.303
P2	1.256	1.897	0.137	0.662	0.576
P1*P2	-0.264	0.404	0.056	-0.653	0.581

The output shows that including the multiplicative term adds only 6/1000 of explained variance to the model that includes only Predictor1. In addition, due to the high correlation between Predictor1 and the multiplicative term and because of the low statistical power, neither parameter for the two predictor terms is statistically significant. The same applies to the entire model. The Tolerance values are particularly low, indicating high variable intercorrelations.

17.6   Regression with Categorical Predictors

Regression with categorical predictors is a frequently addressed issue in the area of applied statistics (Nichols, 1995a, 1995b). Chapter 4 focused on interpretation of slope parameters from categorical predictors. The present section focuses on creating categorical predictors and estimating parameters using SYSTAT.

To illustrate both, we use the data in file "Compstat.sys." First, we illustrate the creation of categorical (also termed subinterval) predictors from continuous predictors. Consider variable Predictor1 in the data file (for the raw data see Section 17.1). For the present purposes we assume that Predictor1 operates at the interval level. To categorize this variable, we specify, for example, two cutoff points. Let the first cutoff point be at Predictor1 = 2.5 and the second at Predictor1 = 4.5. From the two cutoff points we obtain a three-category variable. This variable must be transformed such that pairs of categories result that can be compared using standard OLS regression.

Using SYSTAT, there are several options for performing this transformation. The easiest for very small numbers of cases is to punch in values of one or two new variables. The following routine is more useful when samples assume more realistic sizes. More specifically, we use the Recode option in SYSTAT's worksheet to create new variables that reflect the contrasts we are interested in. For example, we create the following two contrasts:

- $c_1 = (-0.5, -0.5, 1)$. This contrast compares the combined first two categories with the third.

- $c_2 = (-1, 1, 0)$. This contrast compares the first two categories with each other.

Command	Effect
Use Compdat	Reads data file "Compdat.sys"
click Window, Worksheet	Presents data in form of a worksheet
click any variable name and move cursor to the right to the first free field for variable names	Starts a new variable
type variable name, e.g., "Cat2"	Labels first categorical variable Cat2
move one field to the right, type "Cat3"	Initiates new variable; labels it Cat3
to assign values to Cat2 so that they reflect the first contrast, fill the variable and operation boxes as follows:	
If Predictor1 < 4.5 Then Let Cat2 = -0.5; If Predictor1 ≥ 4.5 Then Let Cat2 = 1; OK	Notice that the value 4.5 does not appear for variable Predictor1; had it appeared, this operation would have been equivalent to setting values 4.5 equal to 1
to assign values to Cat3 so that they reflect the second contrast, fill the variable and operation boxes under Recode as follows:	
	continued on next page

If Predictor1 < 2.5 Then Let Cat3 = -1; If Predictor1 > 2.5 Then Let Cat3 = 1; If Predictor1 > 4.5 Then Let Cat3 = 0; OK	
click File, Save	Saves data in file "Compdat.sys"
click Close	Closes worksheet window and readies file "Compdat.sys" for analysis

After these operations, file "Compdat.sys" contains two additional variables, Cat2 and Cat3, which reflect the two contrasts specified above. These variables can be used as predictors in regular multiple regression. The following commands are needed to perform the regression of Criterion on Cat2 and Cat3:

Command	Effect
Read Compdat	Reads data file "Compdat.sys"
click Stats, MGLH, and Regression	Selects regression for data analysis
assign Criterion to Dependent and Cat2 and Cat3 to Independent	Specifies predictors and criterion for multiple regression
click OK	Starts data analysis; sends results to screen

The results from this multiple regression appear in the following output:

```
USE 'A:\COMPDAT.SYS'
SYSTAT Rectangular file A:\COMPDAT.SYS,
created Mon Nov 17, 1997 at 18:20:40, contains variables:
 P1      C       P2      QUAD    LIN     CAT2    CAT3

>MODEL C = CONSTANT+CAT2+CAT3
```

```
>ESTIMATE

Dep Var: C
N: 6
Multiple R: 0.943
Squared multiple R: 0.889

Adjusted squared multiple R: 0.815
Standard error of estimate: 1.414
```

Effect	Coefficient	Std Error	Tol.	t	P(2 Tail)
CONSTANT	8.000	0.577	.	13.856	0.001
CAT2	2.000	0.816	1.0	2.449	0.092
CAT3	3.000	0.707	1.0	4.243	0.024

The output suggests that the two categorical variables allow one to explain $R^2 = 0.89$ of the criterion variance. In addition, the Tolerance values show that the two categorical variables were specified to be orthogonal. They do not share any variance in common and, thus, cover independent portions of the criterion variance.

The following output illustrates this last result. The output displays the results from two simple regression runs, each performed using only one of the categorical predictors.

```
>MODEL C = CONSTANT+CAT2
>ESTIMATE

Dep Var: C
N: 6
Multiple R: 0.471
Squared multiple R: 0.222

Adjusted squared multiple R: 0.028
Standard error of estimate: 3.240
```

Effect	Coefficient	Std Error	Tol.	t	P(2 Tail)
CONSTANT	8.000	1.323	.	6.047	0.004
CAT2	2.000	1.871	1.00	1.069	0.345

```
>MODEL C = CONSTANT+CAT3
>ESTIMATE
```

Dep Var: C
N: 6
Multiple R: 0.816
Squared multiple R: 0.667

Adjusted squared multiple R: 0.583
Standard error of estimate: 2.121

Effect	Coefficient	Std Error	Tol.	t	P(2 Tail)
CONSTANT	8.000	0.866	.	9.238	0.001
CAT3	3.000	1.061	1.00	2.828	0.047

This output indicates that parameter estimates from the two simple regression runs are the same as those from the multiple regression run. In addition, the two R^2 add up to the value obtained from the multiple regression. Independence of results occurs systematically only for orthogonal predictors.

The examples given in this chapter thus far illustrate regression using categorized continuous predictors. When predictors are naturally categorical as, for instance, for the variables gender, make of car, or type of event, the same methods can be applied as illustrated. When the number of categories is two, predictors can be used for regression analysis as they are. When there are three or more categories, one must create contrast variables that compare two sets of variable categories each.

17.7 The Partial Interaction Strategy

Using the means from Chapter 10 (Regression Interaction) and Section 17.6 (Categorizing Variables) we now can analyze hypotheses concerning partial interactions. To evaluate a hypothesis concerning a partial interaction one selects a joint segment of two (or more) variables and tests whether, across this segment, criterion values are higher (or lower) than estimated using the predictor variables' main effects. For each hypothesis of this type there is one parameter. For each parameter there is one coding vector in the design matrix $\mathbf{X}$ of the General Linear Model,

$$\mathbf{y} = \mathbf{X}\mathbf{b} + \mathbf{e}.$$

The coding vectors in $\mathbf{X}$ are created as was illustrated in Section 17.6. Consider the following example. A researcher assumes that criterion values y_i are smaller than expected from the main effects of two predictors, A and B, if the predictors assume values $a_1 < a \le a_2$ and $b_1 < b \le b_2$. To test this hypothesis, we create a coding vector that assumes the following values:

$$c = \begin{cases} -1 & \text{if } a_1 < a \le a_2 \text{ and } b_1 < b \le b_2 \\ 1 & \text{else.} \end{cases}$$

To illustrate this procedure we use our sample data file "Compdat" and first create a categorical variable of the type described in 17.6. Let this variable be named Cat1. We define this variable as follows:

- Cat1 $= -1$ if

    1. Predictor1 is less than 2.5 and Predictor2 is either 1 or 3

    2. Predictor1 is greater than 4.5 and Predictor2 is either 3 or 5

- Cat1 $= 1$ else

Creating a variable according to these specifications proceeds as follows:

Command	Effect
Use Compdat	Reads data file "Compdat.sys"
click Window, Worksheet	Presents data in worksheet form
move cursor to the top line that contains variable names, move to the right in the first free field, type "Cat1"	Creates variable Cat1
move cursor one field down, key "-1," move down, key "-1," move down, key "1," until all values are keyed in	Specifies values for variable Cat1; the values are -1, -1, 1, 1, -1, -1
click File, Save	Saves new and original data in file "Compdat.sys"
click File, Close	Closes Worksheet window and reads file "Compdat.sys"

These commands create the data set displayed in the following output:

```
    P1              C              P2

     1              3               1
     2              5               3
     3              9               2
     4             11               4
     5              9               3
     6             11               5
```

Issuing the following commands, we calculate a regression analysis with partial interaction:

Command	Effect
Use Compdat	Reads data file "Compdat.sys"
click Stats, MGLH, Regression	Selects regression analysis
assign Criterion to Dependent and Predictor1, Predictor2, and Cat1 to Independent	Specifies predictors and criterion for multiple regression
click OK	Starts calculations and sends results to screen

The results from this multiple regression with partial interaction appear in the following output:

```
MODEL C = CONSTANT+P1+P2+CAT1
>ESTIMATE

Dep Var: C
N: 6
Multiple R: 1.000
Squared multiple R: 1.000

Adjusted squared multiple R: 1.000
Standard error of estimate: 0.0

Effect      Coefficient  Std Error  Tol.    t    P(2 Tail)

CONSTANT       2.833        0.0       .      .      .
P1             1.333        0.0      0.309   .      .
P2             0.333        0.0      0.309   .      .
CAT1           1.500        0.0      1.000   .      .
```

To be able to assess the effects of including the partial interaction variable Cat1, we also calculate a multiple regression that involves only

the two main effects of Predictor1 and Predictor2. The results from this
analysis appear in the following output:

```
MODEL C = CONSTANT+P1+P2
>ESTIMATE

Dep Var: C
N: 6
Multiple R: 0.882
Squared multiple R: 0.778

Adjusted squared multiple R: 0.630
Standard error of estimate: 2.000
```

Effect	Coefficient	Std Error	Tolerance	t	P(2 Tail)
CONSTANT	2.333	2.073	.	1.126	0.342
P1	1.333	0.861	0.309	1.549	0.219
P2	0.333	1.139	0.309	0.293	0.789

These two outputs indicate that the portion of criterion variance accounted for by the partial interaction multiple regression model increased
from $R^2 = 0.78$ to $R^2 = 1.00$. Thus, the partial interaction variable
allowed us to explain all of the variance that the two main effect variables were unable to explain. Notice the degree to which Predictor1 and
Predictor2 are dependent upon each other (low Tolerance values).

The first output does not include any statistical significance tests. The
reason for this is that whenever there is no residual variance left, there
is nothing to test against. The F tests (and t tests) test the portion of
variance accounted for against the portion of variance unaccounted for by
the model. If there is nothing left unaccounted for, there is no test.

It is important to note that, in the present example, lack of degrees of
freedom is *not* the reason why there is no statistical test. Including the
partial interaction variable Cat1 implies spending one degree of freedom.
The last output suggests that there are three degrees of freedom to the
residual term and two to the model. Spending one of the residual degrees
of freedom results in a model with three degrees of freedom and in a
residual term with two degrees of freedom. Such a model is testable.

However, a necessary condition is that there is residual variance left.

17.8 Residual Analysis

Analysis of residuals from regression analysis requires the following three steps:

1. Performing regression analysis;

2. Calculating and saving of residuals and estimates of regression diagnostics;

3. Interpretation of estimates of regression diagnostics.

In this section, we use the results from the analysis performed in Section 17.2, where we regressed the variable Criterion onto Predictor1. Results of this regression appeared in the first output, in Section 17.2. The following commands result in this output, and also make the program save estimates, residuals, and indicators for regression diagnostics in a new file:

Command	Effect
Use Compdat	Reads file "Compdat.sys"
click Stats, MGLH	Initiates statistical analysis; selects the MGLH module
click Regression	Selects Regression module
assign Predictor1 to Independent; assign Criterion to Dependent	Specifies predictor and criterion in regression
click Save Residuals	Initiates saving of residuals in user-specified file; program responds by prompting name for file
	continued on next page

| type "Compres" | We name the residual file "Compres" (any other name that meets DOS name specifications would have done); notice that the suffix does not need to be typed; SYSTAT automatically appends ".sys" |
| click OK | Starts calculations; sends results to screen and saves |

The output that results from these commands contains the same information as the earlier output. Therefore, it will not be copied here again. The file "Compres.sys" contains the estimated values, the residuals, and the estimates for regression diagnostics. To be able to compare estimates and residuals with predictor values, we need to merge the files "Compdat.sys" and "Compres.sys." To do this we issue the following commands (to be able to perform the following commands, we need to invoke SYSTAT's command prompt):

Command	Effect
Data	This command carries us into the data module
New	Clears workspace for new data set(s)
Save Compres2	Specifies name for file that will contain merged files, "Compdat.sys" and "Compres.sys"
Use Compdat Compres	Reads the files to be merged; program responds by listing all variables available in the two files
Run	Saves merged data in file "Compres2.sys"; program responds by giving the numbers of cases and variables processed and the name of the merge file

This completes the second step of regression residual analysis, that is, the calculation and saving of residuals, estimates, and regression diagnostics. When interpreting these results, we first inspect the predictor x residual and estimate x residual plots. Second, we check whether any of the regression diagnostics suggest that leverage or distance outliers exist.

To create the plots we first specify output characteristics for the graphs. We issue the following commands:

Command	Effect
Use Compres2	Reads merged file 'Compres2.sys'
click Graph, Options, Global	Opens window for design specifications
click Thickness=3, Decimals to Print = 1, and select from the Graph Font pull down British (or any other font you please); OK	Specifies that lines be printed thick, that one decimal be printed in the axes labels, and that the British type face be used
click Graph, Plot, Plot	Initiates the two-dimensional scatterplot module
assign Predictor1 to X and Criterion to Y	Predictor1 will be on the X-Axis and Criterion will be on the Y-Axis
click Symbol and select 2x for size of data points to print, and the asterisk (or any other symbol you please)	Specifies size and form of data points in print
click Smooth and OK	Creates a linear regression line that coincides with the X-axis
click OK	Sends graph to screen

The graph that results from these commands appears in the left panel of Figure 17.5. Issuing analogous commands we create the right panel of Figure 17.5, which displays the Estimate * Residual Plot.

As is obvious from comparing the two panels of Figure 17.5, the Estimate * Residual plot is very similar to the Predictor * Residual plot.

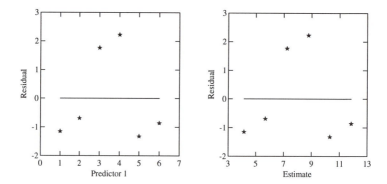

Figure 17.5: Predictor1 * Residual, and Estimate * Residual plots for sample data.

The pattern of data points in the graphs is almost exactly the same. The reason for this great similarity is that the patterns emphasized in both graphs are those that were left unexplained by the regression model. (Figure 17.2 displays the regression of Criterion onto Predictor1. This relationship is statistically significant and allows one to explain 77% of the criterion variance.)

The reason why researchers create Estimate * Residual plots or Predictor * residual plots is that these plots allow one to perform visual searches for systematic patterns of residuals that are associated with the predictor (in simple regression) or the estimates (in simple or multiple regression). While in small samples such patterns may be hard to detect, in the present example it seems that relatively large positive residuals are associated with medium predictor values and with medium estimates. Relatively small negative residuals are associated with the extremes of the predictor and estimate scales.

The following printout displays the file "Compres.sys" that SYSTAT created for the regression of Criterion onto Predictor1:

ESTIMATE	RESIDUAL	LEVERAGE	COOK	STUDENT	SEPRED
4.14	-1.14	0.52	0.48	-0.92	1.27
5.68	-0.68	0.29	0.04	-0.41	0.95
7.22	1.77	0.18	0.13	1.16	0.74
8.77	2.22	0.18	0.21	1.70	0.74

| 10.31 | -1.31 | 0.29 | 0.16 | -0.86 | 0.95 |
| 11.85 | -0.85 | 0.52 | 0.27 | -0.65 | 1.27 |

The output suggests that SYSTAT saves the following variables in its default option: estimate, residual, leverage, the Cook D statistic, student's t, and sepred, that is, the standard error of predicted values. The critical values for the t statistic is $t_{0.05,df=4} = 2.776$. The critical value for the D statistic is $F_{0.05,1,5} = 230$. None of the values in the output are greater than the critical values. Therefore, there is no reason to assume the existence of outliers in this sample data set. Readers are invited to calculate whether cases 6 and 1 exert too much leverage.

The following commands created this output:

Command	Effect
Use Compres	Reads file "Compres.sys"; program responds by listing names of variables in file
click Window, Worksheet	Program transfers data in worksheet
click File, Print	Initiates printing of data file
click OK	Sends entire data file to printer

17.9 Missing Data Estimation

As by the date of this writing, SYSTAT does not provide a separate module for missing data estimation. Only in the time series programs missing values can be interpolated using distance weighted least squares methods. There are modules for missing data estimation in BMDP and EQS. The EQS module provides the options to estimate and impute variable means, group means, and values estimated by multiple regression. In this section we focus on estimating and imputing missing data using regression methods from SYSTAT's MGLH regression module. We focus on data with missing criterion values. Data with missing predictor values can be treated accordingly.

This module allows one to estimate missing data from a regression model. Estimation and imputation of missing values requires three steps:

1. Estimate regression model;

2. Save regression estimates, residuals, and diagnostics using the default option (Section 17.6);

3. Impute estimates for missing values into original data file; the residuals file contains these estimates.

To be able to use the data in file "Compdat.sys," we need to add at least one case with a missing criterion value. We add the values Predictor1 = 12 and Predictor2 = 14. To do this in SYSTAT, we perform the following steps:

Command	Effect
Use Compdat	Reads data file 'Compdat.sys'
click Window, Worksheet	Transfers data file into worksheet window
move cursor down to first line below last data line, move to Predictor1 column, key in "12"	Specifies Predictor1 value for new case
move to Predictor2 column, key in "14"	Specifies Predictor2 value for new case
click File, Save As	Initiates saving of completed data in new file
type "Compdat2"; OK	Saves completed data in file "Compdat2.sys"; leaves all other values for new case missing; missing values are indicated by a period
click File, Close	Carries us back to command mode where we can perform statistical analyses

For the following analyses we only use the three variables Predictor1,

Predictor2, and Criterion. The values assumed by these three variables appear below.

P1	C	P2
1	3	1
2	5	3
3	9	2
4	11	4
5	9	3
6	11	5
12	.	14

The output displays the values the three variables assume for the original six cases. In addition, the values Predictor1 = 12 and Predictor2 = 14 appear, along with the missing value for Criterion. This selective Output was created within SYSTAT's command mode, issuing the following commands (these commands assume that file 'Compdat.sys' is the active file; if this is not the case, it must be opened before issuing the following commands):

Command	Effect
Caselist Pred1, Pred2, Crit	Initiates listing of values of variables Pred1, Pred2, and Crit; all cases will be displayed
Run	Displays all cases on screen
highlight desired parts of display, click File, Print, OK	Sends highlighted parts of what is displayed on screen to printer

Using file "Compdat2.sys" we now estimate the parameters for the multiple regression of Criterion onto Predictor1 and Predictor2. The following output displays results from this analysis. The commands for this run are analogous to the commands for regression with the saving of the residual file and are not given here.

MODEL C = CONSTANT+P1+P2

```
>ESTIMATE
1 case(s) deleted due to missing data.

Dep Var: C
N: 6
Multiple R: 0.882
Squared multiple R: 0.778

Adjusted squared multiple R: 0.630
Standard error of estimate: 2.000

Effect      Coefficient   Std Error    t      P(2 Tail)

CONSTANT       2.333        2.073     1.126    0.342
P1             1.333        0.861     1.549    0.219
P2             0.333        1.139     0.293    0.789
```

With only three exceptions, this output is identical with the output
on p. 316. For the present purposes most important is that the case with
the missing criterion value was excluded from analysis. This is indicated
in the line before the numerical results block. The last line of the above
output (not given here) indicates that the residuals have been saved.

File "Cd2res.sys" contains results from residual analysis. Using the
same commands as before, we print this file. It appears in the following
output:

```
ESTIMATE   RESIDUAL  LEVERAGE   COOK   STUDENT   SEPRED
    4         -1        0.58     0.28   -0.70      1.52
    6         -1        0.58     0.28   -0.70      1.52
    7          2        0.33     0.25    1.41      1.15
    9          2        0.33     0.25    1.41      1.15
   10         -1        0.58     0.28   -0.70      1.52
   12         -1        0.58     0.28   -0.70      1.52
   23          .       14.66      .       .        7.65
```

This output displays the same type of information as before. In addition, it contains information of importance for the purposes of estimating
and imputing missing data. For the last case, that is, the case with the

missing Criterion value, this file shows an estimate, a leverage value, and the standard error of the predicted value. It is the estimate that we impute into the original file. To perform this we "Use" data file "Compdat2.sys," transfer the data to the worksheet window, and impute 23.0 for the missing Criterion value. Using the now completed data we recalculate the regression analysis. Results of this analysis appear in following output:

```
MODEL C = CONSTANT+P1+P2
>ESTIMATE

Dep Var: C
N: 7
Multiple R: 0.975
Squared multiple R: 0.951

Adjusted squared multiple R: 0.927
Standard error of estimate: 1.732
```

Effect	Coefficient	Std Error	t	P(2 Tail)
CONSTANT	2.333	1.224	1.906	0.129
P1	1.333	0.686	1.944	0.124
P2	0.333	0.573	0.581	0.592

Analysis of Variance

Source	SS	df	MS	F	P
Regression	234.857	2	117.429	39.143	0.002
Residual	12.000	4	3.000		

This output shows a number of interesting results. First, the number of cases processed now is $n = 7$ rather than $n = 6$ as in the earlier regression output. No case was eliminated because of missing data. Second, the three parameter estimates are exactly as before. The reason is that the estimated value sits exactly on the regression hyperplane. Therefore, the new data point changes neither the elevation nor the angles of the hyperplane.

Also because the new data point sits exactly on the regression plane, the overall amount of variance that is unaccounted for remains the same. Thus, considering the larger number of data points processed, the portion of variance accounted for is greater than that without the imputed data point. This can be seen in the analysis of variance panel. As a result, the standard errors of the parameters are smaller than before, and so are the tail probabilities of the t statistics. Overall, the regression model now accounts for a statistically significant portion of the criterion variance.

In general, one can conclude that good estimation and imputation of missing values tends to increase statistical power. The main reasons for this increase include the increase in sample size that results from including cases with imputed data and the placing of OLS estimates on the regression hyperplane.

17.10   Piecewise Regression

SYSTAT provides two options for estimating continuous piecewise regression models. The first is to regress a criterion variable onto vectors from a user-specified design matrix of the form given in (15.2) in Section 15.1. The second is to estimate a model from a user-specified form. SYSTAT's Nonlin module allows one to perform the second type of analysis. This section focuses on the second option.

To illustrate piecewise regression we use the file "Compdat.sys" again. We analyze the relationship between the variables Predictor1 and Criterion. Figure 17.2 suggests that there may be a change in the relationship between Predictor1 and Criterion at Predictor1 = 3.5. Therefore, we estimate a piecewise regression model of the following form (cf. Section 15.1):

$$\text{Criterion} = \text{Constant} + b_1 * \text{Predictor1}$$
$$+ b_2 * (\text{Predictor2} - 3.5) + \text{Residual}.$$

Using SYSTAT, this model can be estimated issuing the following commands:

Command	Effect
Use Compdat	Reads file 'Compdat.sys'
click Stats, Nonlin	Invokes Nonlin module for nonlinear estimation of regression models
click Model	Opens window for model specification; the Loss Function option does not need to be invoked unless one wishes to specify a goal function other than least squares
type into the model window "crit = constant + b1*pred1 + b2*(pred1 - 3.5)"	Specifies piecewise regression function to be fit
click OK	Starts estimation; sends results to screen

The following output displays results from the estimation process:

```
MODEL c = constant+b1*p1+b2*(p1-3.5)

>ESTIMATE / QUASI

Iteration
No.     Loss      CONSTANT    B1          B2
  0   .86985D+02  .1011D+01  .1020D+01   .1030D+01
  1   .37150D+02  .1240D+01  .1724D+01   .9311D+00
  2   .12343D+02  .1617D+01  .1825D+01  -.2868D+00
  3   .12343D+02  .1615D+01  .1824D+01  -.2814D+00

Dependent variable is C

     Source   Sum-of-Squares   df   Mean-Square
 Regression       425.657       3      141.886
   Residual        12.343       3        4.114

      Total       438.000       6
```

```
Mean corrected        54.000     5
```

```
        Raw  R-square (1-Residual/Total)        = 0.972
Mean corrected R-square (1-Residual/Corrected) = 0.771
        R(observed vs predicted) square        = 0.771
```

```
                                    Wald Conf. Interval
Parameter  Estimate  A.S.E.    Lower < 95%> Upper
CONSTANT    1.615    51479   -163828.536    163831.766
B1          1.824    14708    -46806.790     46810.439
B2         -0.281    14708    -46808.896     46808.333
```

The first line of this output shows the model specification. The second line shows the Estimate command and, after the slash, the word QUASI. This word indicates that the program employs a quasi-Newton algorithm for estimation. What follows is a report of the iteration process. Information provided includes the number of iteration, the value of the loss function, and parameter values for each iteration step. The report suggests that convergence was achieved after the fourth iteration step. The last part of the output displays results of the statistical analysis of the regression model, the raw R^2 and the corrected R^2, and the parameter estimates.

Dividing the regression sum of squares by the residual sum of squares yields the F ratio for the regression model. For the present example we obtain $F = 141.886/4.114 = 34.489$. This value is, for $df_1 = 3$ and $df_2 = 3$ greater than the critical value $F_{0.05,3,3} = 9.28$. We thus conclude that the regression model accounts for a statistically significant portion of the criterion variance.

Wilkinson et al. (1992, p. 429) suggest you "never trust the output of an iterative nonlinear estimation procedure until you have plotted estimates against predictors ..." Therefore, we now create a plot of the data from file "Compdat.sys" in which we lay the piecewise regression line along with the standard linear regression line. Figure 17.6 displays this plot.

The straight line in this plot represents the standard linear regression line for the regression of Criterion onto Predictor1. The second line represents the piecewise regression line. The two lines coincide for the segment where Predictor1 is greater than 3.5. For the segment where Predictor1 is less than or equal to 3.5, the piecewise regression line has a steeper slope, thus going almost exactly through the two data points for Predictor1 = 2 and Predictor1 = 2.

The following commands create this plot:

Command	Effect
Use Compdat	Reads data file "Compdat.sys"
click Graph, Option	Opens window for specification of general graph characteristics
select Thickness = 3, Decimals to Print = 1, and Graph Font = British	Specifies thickness of lines, number of decimals, and typeface
click OK	Carries us back to command mode
click Graph, Function	Opens window that allows one to specify function to be plotted
type "y = 0.83 + 2.049 * x - 0.506 * (x - 3.5) * (x > 3.5)"	Specifies function to be plotted
click Axes	Opens window for specification of axes' characteristics
insert xmin = 0, ymin = 2, xmax = 7, ymax = 12, xlabel = Pred1, and ylabel = Crit; click OK	Specifies minimum and maximum axes values and labels axes
click OK, OK	Sends graph to screen
click within the Graph Window the Window, Graph Placement, and Overlay Graph options	Tells the program that a second graph will follow that is to be laid over the first
click Graph, Plot, Plot	Invokes module for plotting data in 2D
assign Predictor1 to the X axis and Criterion to the Y axis	
	continued on next page

click Symbol and select *, Size = 2, OK	Determines size and type of symbol to use for depicting data points
click Smooth, OK	Tells the program to insert a linear regression line into the data plot; the linear smoothing is default
click Axes and specify xmin = 0, ymin = 2, xmax = 7, ymax = 12, xlabel = Pred1, and ylabel = Crit; OK	Specifies min and max values for axes and axes labels; this is needed for the same reason as the title
OK	Redraws and overlays both graphs; sends picture to screen

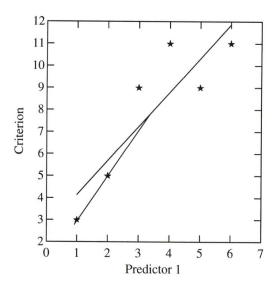

Figure 17.6: Comparing straight line and piecewise regression.

Appendix A

ELEMENTS OF MATRIX ALGEBRA

This excursus provides a brief introduction to elements of matrix algebra. More detailed introductions that also cover more ground can be found in such textbooks as Ayres (1962), Searle (1982), Graybill (1983). This excursus covers the following elements of matrix algebra: definition of a matrix, types of matrices, transposing a matrix, addition and subtraction with matrices, multiplication with matrices, the rank of a matrix, the inverse of a matrix, and the determinant of a matrix.

A.1 Definition of a Matrix

Definition: A matrix is a rectangular array of numbers. These numbers are arranged in rows and columns. Consider matrix $\mathbf{A}$ with m rows and n columns, that is, the $m \times n$ matrix $\mathbf{A}$. The numbers in this matrix are arranged as

$$
\mathbf{A} = \begin{pmatrix} a_{11} & a_{12} & \dots & a_{1n} \\ a_{21} & a_{22} & \dots & a_{2n} \\ \vdots & \vdots & \ddots & \vdots \\ a_{m1} & a_{m2} & \dots & a_{mn} \end{pmatrix}, \tag{A.1}
$$

with $i = 1, \ldots, m$, and $j = 1, \ldots, n$.

The presentation of a matrix such as given in (A.1) meets the following conventions:

1. Matrices are denoted using capital letters, for example, $\mathbf{A}$.

2. The a_{ij} are the numbers or elements of matrix $\mathbf{A}$. Each number is placed in a cell of the matrix; subscripts denote cell indexes.

3. The subscripts, ij, give first the row number, i, and second the column number, j. For example, subscript 21 indexes the cell that is in the second row in the first column.

4. Matrix elements, that is, cells, are denoted using subscripted lower case letters, for example, a_{ij}.

The following example presents a 2 x 3 matrix with its elements:

$$\mathbf{A} = \left(\begin{array}{ccc} a_{11} & a_{12} & a_{13} \\ a_{21} & a_{22} & a_{23} \end{array} \right).$$

Matrices can be considered mathematical objects just as, for instance, real valued numbers. Therefore, it is natural to ask how one can define such operations as addition or multiplication. Are there neutral and inverse elements for addition? Are there neutral and inverse elements for multiplication? Generally, the answer is yes, as long as addition is concerned. However, when it comes to multiplication, an inverse element does not always exist. This may not be intuitively plausible, since for real numbers there is for every number a an element $1/a$ which is the inverse of a. It is also true that addition and multiplication are not well-defined for arbitrary matrices. This is also not the case for real numbers. Usually, any two real numbers can be added or multiplied. Matrix multiplication is generally not even commutative, that is, if $\mathbf{A}$ and $\mathbf{B}$ are two matrices that can be multiplied with each other, then generally $\mathbf{AB} \neq \mathbf{BA}$. However, the situation is not as discouraging as it may seem at first glance. One gets rapidly familiar with matrix operations. The following sections introduce readers into matrix definitions and operations needed in subsequent chapters.

A.2 Types of Matrices

This section presents the following types of matrices: square matrix, column vector, row vector, diagonal matrix, identity matrix, and triangular forms of matrices.

1. *Square matrices.* Matrices are square if the number of rows equals the number of columns. For example, matrices with dimensions 2 x 2 or 5 x 5 are square. The following is an example of a 2 x 2 matrix:

$$\mathbf{B} = \left(\begin{array}{cc} 12.31 & -4.45 \\ -36.02 & 0.71 \end{array} \right).$$

2. *Column vectors.* A matrix with only one column is termed a column vector. For example, the following are sample dimensions of column vectors: 7 x 1, 3 x 1, and 2 x 1. The following is an example of a 3 x 1 column vector:

$$\mathbf{a} = \left(\begin{array}{c} a_1 \\ a_2 \\ a_3 \end{array} \right).$$

Since there is only one column a subscript indicating the column is redundant and therefore omitted.

3. *Row vectors.* A matrix with only one row is termed a row vector. The following are sample dimensions of row vectors: 1 x 2, 1 x 45, and 1 x 3. The following is an example of a row vector

$$\mathbf{a}' = \left(\begin{array}{ccc} 114.1 & -32.8 & -1.9 \end{array} \right).$$

For the remainder of this volume we adopt the following convention: Whenever we speak of vectors we always mean column vectors. Row vectors will be denoted by a prime.[1] For example, if vector $\mathbf{a}$ is a row vector, we express it as $\mathbf{a}'$.

4. *Diagonal matrices.* A square matrix with numbers in its diagonal cells and zeros in its off-diagonal cells is termed a diagonal matrix.

---

[1] See the definition of a transposed matrix below.

The elements $a_{11}, a_{22}, \ldots$, describe the main diagonal of matrix $\mathbf{A}$; that is, the main diagonal of a matrix is constituted by the cells with equal indexes. For example, the diagonal cells of a 3 x 3 matrix have indexes 11, 22, and 33. These are the cells that go from the upper left corner to the lower right corner of a matrix. When a matrix is referred to as diagonal, reference is made to a matrix with a main diagonal. The following is an example of a 3 x 3 diagonal matrix:

$$\mathbf{A} = \begin{pmatrix} a_{11} & 0 & 0 \\ 0 & a_{22} & 0 \\ 0 & 0 & a_{33} \end{pmatrix}.$$

Usually, diagonal matrices are written as

$$\mathbf{A} = diag\begin{pmatrix} a_{11} & a_{22} & a_{33} \end{pmatrix}.$$

If the elements a_{ii} are matrices themselves $\mathbf{A}$ is termed a *block diagonal* matrix.

5. *Identity matrices.* A diagonal matrix with diagonal elements that are equal to some constant value k, that is, $a_{11} = a_{22} = a_{33} = \ldots = a_{nn} = k$, is called a *scalar matrix*. If, in addition, the constant $k = 1$, the matrix is termed an *identity matrix*. The symbol for an $n \times n$ identity matrix is $\mathbf{I_n}$, for example,

$$\mathbf{I_3} = \begin{pmatrix} 1 & 0 & 0 \\ 0 & 1 & 0 \\ 0 & 0 & 1 \end{pmatrix}.$$

6. *Triangular matrices.* A square matrix with elements $a_{ij} = 0$ for $j < i$ is termed an *upper triangular matrix*. The following is an example of a 2 x 2 upper triangular matrix:

$$\mathbf{U} = \begin{pmatrix} a_{11} & a_{12} \\ 0 & a_{22} \end{pmatrix}.$$

A square matrix with elements $a_{ij} = 0$ for $j > i$ is termed a *lower triangular matrix*. The following is an example of a 3 x 3 lower triangular matrix:

$$\mathbf{L} = \begin{pmatrix} a_{11} & 0 & 0 \\ a_{21} & a_{22} & 0 \\ a_{31} & a_{32} & a_{33} \end{pmatrix}.$$

Diagonal matrices are both upper and lower triangular.

A.3 Transposing Matrices

Consider the $m \times n$ matrix $\mathbf{A}$. Interchanging rows and columns of $\mathbf{A}$ yields $\mathbf{A}'$, the transpose[2] of $\mathbf{A}$. By transposing $\mathbf{A}$, one moves cell ij to be cell ji. The following example transposes the 2 x 3 matrix, $\mathbf{A}$:

$$\mathbf{A} = \begin{pmatrix} 1 & 2 & 3 \\ 4 & 5 & 6 \end{pmatrix} \rightarrow \mathbf{A}' = \begin{pmatrix} 1 & 4 \\ 2 & 5 \\ 3 & 6 \end{pmatrix}. \tag{A.2}$$

As (A.2) shows, after transposition, what used to be rows are columns, and what used to be columns are rows. Transposing a transposed matrix yields the original matrix, or, more specifically, $\mathbf{A}'' = \mathbf{A}$. Matrices for which $\mathbf{A}' = \mathbf{A}$ are termed symmetric.

A.4 Adding Matrices

Adding matrices is only possible if the dimensions of the matrices are the same. For instance, adding $\mathbf{A}$ with dimensions m and n and $\mathbf{B}$ with dimensions p and q can be performed only if both $m = p$ and $n = q$. This is the case in the following example. If $\mathbf{A}$ and $\mathbf{B}$ are given by

$$\mathbf{A} = \begin{pmatrix} 1 & 5 \\ 10 & 3 \end{pmatrix} \quad \text{and} \quad \mathbf{B} = \begin{pmatrix} 4 & 3 \\ 2 & 5 \end{pmatrix}.$$

---

[2]Instead of the prime symbol one also finds in the literature the symbol $\mathbf{A}^{\mathrm{T}}$ to denote a transpose.

The sum of these two matrices is calculated as

$$\mathbf{A} + \mathbf{B} = \begin{pmatrix} 1+4 & 5+3 \\ 10+2 & 3+5 \end{pmatrix} = \begin{pmatrix} 5 & 8 \\ 12 & 8 \end{pmatrix}. \qquad (A.3)$$

Rather than giving a formal definition of matrix addition, this example shows that the corresponding elements of the two matrices are simply added to create the element of the new matrix $\mathbf{A} + \mathbf{B}$.

In contrast, matrix $\mathbf{A}$ with dimensions 3 x 6 and matrix $\mathbf{B}$ with dimensions 3 x 3 cannot be added to each other. In a fashion analogous to (A.3) one can subtract matrices from each other, if they have the same dimensions.

A.5 Multiplying Matrices

This excursus covers three aspects of matrix multiplication:

1. Multiplication of a matrix with a scalar;

2. Multiplication of two matrices with each other; and

3. Multiplication of two vectors.

Multiplication of a matrix with a scalar is performed by multiplying each of its elements with the scalar. For example, multiplication of scalar $k = 3$ with matrix $\mathbf{A}$ from Section A.4 yields

$$k\mathbf{A} = 3 \begin{pmatrix} 1 & 5 \\ 10 & 3 \end{pmatrix} = \begin{pmatrix} 3 & 15 \\ 30 & 9 \end{pmatrix}$$

It should be noted that multiplication of a matrix with a scalar is commutative, that is, $k\mathbf{A} = \mathbf{A}k$ holds.

Multiplication of two matrices

Matrices must possess one specific characteristic to be multipliable with each other. Consider the two matrices, $\mathbf{A}$ and $\mathbf{B}$. Researchers wish to multiply them to calculate the matrix product $\mathbf{AB}$. This is possible only if the number of columns of $\mathbf{A}$ is equal to the number of rows of $\mathbf{B}$.

One calls **A**, that is, the first of two multiplied matrices, postmultiplied by **B**, and **B**, that is, the second of two multiplied matrices, premultiplied by **A**. In other words, two matrices can be multiplied with each other if the number of columns of the postmultiplied matrix is equal to the number of rows of the premultiplied matrix.

When multiplying two matrices one follows the following procedure: one multiplies row by column, and each element of the row is multiplied by the corresponding element of the column. The resulting products are summed. This sum of products is one of the elements of the resulting matrix with the row index carried over from the row of the postmultiplied matrix and the column index carried over from the column of the premultiplied matrix. Again, the number of columns of the postmultiplied matrix must be equal to the number of rows of the premultiplied matrix.

For example, to be able to postmultiply **A** with **B**, **A** must have dimensions $m \times n$ and **B** must have dimensions $n \times p$. The resulting matrix has the number of rows of the postmultiplied matrix and the number of columns of the premultiplied matrix. In the present example, the product **AB** has dimensions $m \times p$. The same applies accordingly when multiplying vectors with each other or when multiplying matrices with vectors (see below).

Consider the following example. A researcher wishes to postmultiply matrix **A** with matrix **B** (which is the same as saying the researcher wishes to premultiply matrix **B** with matrix **A**). Matrix **A** has dimensions $m \times n$. Matrix **B** has dimensions $p \times q$. **A** can be postmultiplied with **B** only if $n = p$. Suppose matrix **A** has dimensions 3 x 2 and matrix **B** has dimensions 2 x 2. Then, the product **AB** of the two matrices can be calculated using the following multiplication procedure. The matrices are

$$
\mathbf{A} = \begin{pmatrix} a_{11} & a_{12} \\ a_{21} & a_{22} \\ a_{31} & a_{32} \end{pmatrix} \quad \text{and} \quad \mathbf{B} = \begin{pmatrix} b_{11} & b_{12} \\ b_{21} & b_{22} \end{pmatrix}.
$$

The matrix product **AB** is given in general form by

$$
\mathbf{AB} = \begin{pmatrix} a_{11}b_{11} + a_{12}b_{21} & a_{11}b_{12} + a_{12}b_{22} \\ a_{21}b_{11} + a_{22}b_{21} & a_{21}b_{12} + a_{22}b_{22} \\ a_{31}b_{11} + a_{32}b_{21} & a_{31}b_{12} + a_{32}b_{22} \end{pmatrix}.
$$

For a numerical example consider the following two matrices:

$$\mathbf{A} = \begin{pmatrix} 2 & 3 \\ 4 & 1 \\ 3 & 5 \end{pmatrix} \quad \text{and} \quad \mathbf{B} = \begin{pmatrix} 3 & 4 \\ 7 & 8 \end{pmatrix}.$$

Their product yields

$$\mathbf{AB} = \begin{pmatrix} 2*3+3*7 & 2*4+3*8 \\ 4*3+1*7 & 4*4+1*8 \\ 3*3+5*7 & 3*4+5*8 \end{pmatrix} = \begin{pmatrix} 27 & 32 \\ 19 & 24 \\ 44 & 52 \end{pmatrix}.$$

As vectors are defined as a matrix having only one column, there is nothing special in writing, for instance, the product of a matrix and a vector $\mathbf{Ab}$ (yielding a column vector), or the product of a row vector with a matrix $\mathbf{b'A}$ (yielding a row vector). As a matter of course, matrix $\mathbf{A}$ and vector $\mathbf{b}$ must have appropriate dimensions for these operations, or, in more technical terms, $\mathbf{A}$ and $\mathbf{b}$ must be conformable to multiplication.

Multiplication of two vectors with each other

Everything that was said concerning the multiplication of matrices carries over to the multiplication of two vectors $\mathbf{a}$ and $\mathbf{b}$, with no change. So, no extra rules need to be memorized. While the multiplication of two column vectors and the multiplication of two row vectors is impossible according to the definition for matrix multiplication, the product of a row vector with a column vector and the product of a column vector with a row vector are possible. These two products are given special names. The product of a row vector with a column vector, that is, $\mathbf{a'b}$, is called the *inner product*. Alternatively, because the result is a scalar, this product is also termed the scalar product. The scalar product is sometimes denoted by $< \mathbf{a}, \mathbf{b} >$.

In order to calculate the inner product both vectors must have the same number of elements. Consider the following example of the two three-element vectors **a** and **b**. Multiplication of

$$\mathbf{a'} = \begin{pmatrix} a_{11} & a_{12} & a_{13} \end{pmatrix} \quad \text{with} \quad \mathbf{b} = \begin{pmatrix} b_{11} \\ b_{21} \\ b_{31} \end{pmatrix}$$

yields the inner product

$$\mathbf{a'b} = a_{11}b_{11} + a_{12}b_{21} + a_{13}b_{31} = \sum_{i=1}^{3} a_{1i}b_{i1}.$$

For a numerical example consider the vectors

$$\mathbf{a'} = \begin{pmatrix} 3 & 6 & 1 \end{pmatrix} \quad \text{and} \quad \mathbf{b} = \begin{pmatrix} 2 \\ 7 \\ 9 \end{pmatrix}.$$

The inner product is then calculated as

$$\mathbf{a'b} = 3*2 + 6*7 + 1*9 = 57.$$

Most important for the discussion of multicollinearity in multiple regression (and analysis of variance) is the concept of orthogonality of vectors. Two vectors are orthogonal if their inner product is zero. Consider the two vectors a' and b. These two vectors are orthogonal if

$$\sum_{i} a_i b_i = \mathbf{a'b} = 0.$$

The product of a column vector with a row vector **ab'** is called the *outer product* and yields a matrix with a number of rows equal to the number of elements of **a** and number of columns equal to the number of elements of **b**. Transposing both **a'** and **b** yields **a** and **b'**. **a** has dimensions 3 x 1 and **b'** has dimensions 1 x 3. Multiplying

$$\mathbf{a} = \begin{pmatrix} a_{11} \\ a_{21} \\ a_{31} \end{pmatrix} \quad \text{with} \quad \mathbf{b'} = \begin{pmatrix} b_{11} & b_{12} & b_{13} \end{pmatrix}$$

yields the outer product

$$\mathbf{ab}' = \begin{pmatrix} a_{11}b_{11} & a_{11}b_{12} & a_{11}b_{13} \\ a_{21}b_{11} & a_{21}b_{12} & a_{21}b_{13} \\ a_{31}b_{11} & a_{31}b_{12} & a_{31}b_{13} \end{pmatrix}.$$

Using the numbers of the last numeric example yields

$$\mathbf{ab}' = \begin{pmatrix} 3*2 & 3*7 & 3*9 \\ 6*2 & 6*7 & 6*9 \\ 1*2 & 1*7 & 1*9 \end{pmatrix} = \begin{pmatrix} 6 & 21 & 27 \\ 12 & 42 & 54 \\ 2 & 7 & 9 \end{pmatrix}.$$

Obviously, the order of factors is significant when multiplying vectors with each other. In the present context, the inner product is more important.

A.6   The Rank of a Matrix

Before introducing readers to the concept of rank of a matrix, we need to introduce the concept of linear dependency of rows or columns of a matrix. Linear dependency can be defined as follows: The columns of a matrix $\mathbf{A}$ are linearly dependent if there exists a vector $\mathbf{x}$ such that $\mathbf{Ax}$ yields a vector containing only zeros. In other words, if the equation $\mathbf{Ax} = \mathbf{0}$ can be solved for $\mathbf{x}$ then the columns of $\mathbf{A}$ are linearly dependent. Because this question would be trivial if we allow $\mathbf{x}$ to be a vector of zeros we exclude this possibility. Whether such a vector exists can be checked by solving the equation $\mathbf{Ax} = \mathbf{0}$ using, for instance, the Gauß Algorithm. If an $\mathbf{x}$ other than the trivial solution $\mathbf{x} = \mathbf{0}$ exists, the columns are said to be linearly dependent. Consider, for example, the following 3 x 3 matrix:

$$\mathbf{A} = \begin{pmatrix} 1 & 2 & 3 \\ -2 & -4 & -6 \\ -3 & -6 & 5 \end{pmatrix}.$$

If we multiply $\mathbf{A}$ by the column vector $\mathbf{x}' = (2, -1, 0)$ we obtain $\mathbf{b}' = (0, 0, 0)$. Vector $\mathbf{x}$ is not an array of zeros. Therefore the columns of $\mathbf{A}$ are said to be linearly dependent.

A relatively simple method for determining whether rows or columns of a matrix are linearly independent involves application of such linear operations as addition/subtraction and multiplication/division. Consider the following example of a 3 x 3 matrix:

$$\mathbf{A} = \begin{pmatrix} 1 & -2 & -3 \\ -4 & 2 & 0 \\ 5 & -3 & -1 \end{pmatrix}.$$

The columns of this matrix are linearly dependent. The following operations yield a row vector with only zero elements:

1. Multiply the second column by two; this yields

$$\begin{pmatrix} -4 \\ 4 \\ -6 \end{pmatrix}.$$

2. Add the result of Step 1 to the first column; this yields

$$\begin{pmatrix} -3 \\ 0 \\ -1 \end{pmatrix}.$$

3. Subtract the third column from the result obtained in the second step; this yields a column vector of zeros. Thus, the columns are linearly dependent.

If the columns of a matrix are linearly independent, the rows are independent also, and vice versa. Indeed, it is one of the interesting results of matrix algebra that the number of linearly independent columns of a matrix always equals the number of linearly independent rows of the matrix.

We now turn to the question of how many linear dependencies there are among the columns of a matrix. This topic is closely related to the rank of a matrix.

To introduce the concept of rank of a matrix, consider matrix $\mathbf{A}$ with dimensions $m \times n$. The rank of a matrix is defined as the number of linearly independent rows of this matrix. If the rank of a matrix equals the number of columns, that is, $rank(\mathbf{A}) = n$, the matrix has full column rank. Accordingly, if $rank(\mathbf{A}) = m$, the matrix has full row rank. If a square matrix has full column rank (and, therefore, full row rank as well), it is said to be nonsingular. In this case, the inverse of this matrix exists (see the following section). For numerical characteristics of the rank of a matrix we refer readers to textbooks for matrix algebra (e.g., Ayres, 1962).

A.7   The Inverse of a Matrix

There is no direct way of performing divisions of matrices. One uses inverses of matrices instead. To explain the concept of an inverse consider the two matrices $\mathbf{A}$ and $\mathbf{B}$. Suppose we postmultiply $\mathbf{A}$ with $\mathbf{B}$ and obtain $\mathbf{AB} = \mathbf{I}$, where $\mathbf{I}$ is the identity matrix. Then, we call $\mathbf{B}$ the inverse of $\mathbf{A}$. Usually, inverse matrices are identified by the superscript "-1," Therefore, we can rewrite the present example as follows: $\mathbf{B} = \mathbf{A}^{-1}$. It also holds that both pre- and postmultiplication result in $\mathbf{I}$, that is,

$$\mathbf{AA}^{-1} = \mathbf{A}^{-1}\mathbf{A} = \mathbf{I}. \tag{A.4}$$

For an inverse to exist, a matrix must be nonsingular, square (although not all square matrices have an inverse), and of full rank. The inverse of a matrix can be viewed similarly to the reciprocal of a number in ordinary linear algebra. Consider the number 2. Its reciprocal is 1/2. Multiplying 2 by its reciprocal, 1/2, gives 1. In general, a number multiplied with its reciprocal always equals 1,

$$x\frac{1}{x} = 1.$$

Reexpressing this in a fashion parallel to (A.4) we can write

$$x\frac{1}{x} = xx^{-1} = 1.$$

Calculating an inverse for a matrix can require considerable amounts of computing. Therefore, we do not provide the specific procedural steps for calculating an inverse in general. All major statistical software packages include modules that calculate inverses of matrices. However, we do give examples for two special cases, for which inverses are easily calculated. These examples are the inverses of diagonal matrices and inverses of matrices of 2 x 2 matrices. The inverse of a diagonal matrix is determined by calculating the reciprocal values of its diagonal elements. Consider the 3 x 3 diagonal matrix

$$\mathbf{A} = \begin{pmatrix} a_{11} & 0 & 0 \\ 0 & a_{22} & 0 \\ 0 & 0 & a_{33} \end{pmatrix}.$$

The inverse of this matrix is

$$\mathbf{A}^{-1} = \begin{pmatrix} 1/a_{11} & 0 & 0 \\ 0 & 1/a_{22} & 0 \\ 0 & 0 & 1/a_{33} \end{pmatrix}.$$

This can be verified by multiplying $\mathbf{A}$ with $\mathbf{A}^{-1}$. The result will be the identity matrix $\mathbf{I_3}$.

Consider the following numerical example of a 2 x 2 diagonal matrix $\mathbf{A}$ and its inverse $\mathbf{A}^{-1}$:

$$\mathbf{A} = \begin{pmatrix} 3 & 0 \\ 0 & 6 \end{pmatrix}, \quad \mathbf{A}^{-1} = \begin{pmatrix} 1/3 & 0 \\ 0 & 1/6 \end{pmatrix}. \tag{A.5}$$

Multiplying $\mathbf{A}$ with $\mathbf{A}^{-1}$ results in

$$\mathbf{A}\mathbf{A}^{-1} = \begin{pmatrix} 3*1/3+0*0 & 3*0+0*1/6 \\ 0*1/3+6*0 & 0*0+6*1/6 \end{pmatrix} = \begin{pmatrix} 1 & 0 \\ 0 & 1 \end{pmatrix} = \mathbf{I_2},$$

which illustrates that multiplying a matrix by its inverse yields an identity matrix.

The inverse of a 2 x 2 matrix, $\mathbf{A}^{-1}$, can be calculated as

$$\mathbf{A}^{-1} = \frac{1}{a_{11}a_{22} - a_{12}a_{21}} \begin{pmatrix} a_{22} & -a_{12} \\ -a_{21} & a_{11} \end{pmatrix},$$

or, after setting $D = a_{11}a_{22} - a_{12}a_{21}$ we obtain

$$\mathbf{A}^{-1} = \frac{1}{D} \begin{pmatrix} a_{22} & -a_{12} \\ -a_{21} & a_{11} \end{pmatrix}. \tag{A.6}$$

To illustrate this way of calculating the inverse of a matrix we use the same matrix as in (A.5). We calculate $D = 3*6 - 0*0 = 18$. Inserting into (A.6) yields

$$\mathbf{A}^{-1} = \frac{1}{18} \begin{pmatrix} 6 & -0 \\ -0 & 3 \end{pmatrix} = \begin{pmatrix} 1/3 & 0 \\ 0 & 1/6 \end{pmatrix}. \tag{A.7}$$

Obviously, (A.5) is identical to (A.7). The inverses of matrices are needed to replace algebraic division. This is most important when solving equations. Consider the following example. We have the matrix equation $\mathbf{A} = \mathbf{BY}$ that we wish to solve for $\mathbf{Y}$. Suppose that the inverse of $\mathbf{B}$ exists. We perform the following steps:

1. Premultiply both sides of the equation with $\mathbf{B}^{-1}$. We then obtain the expression $\mathbf{B}^{-1}\mathbf{A} = \mathbf{B}^{-1}\mathbf{BY}$.

2. Because of $\mathbf{B}^{-1}\mathbf{BY} = \mathbf{IY} = \mathbf{Y}$ we have a solution for the equation. It is $\mathbf{Y} = \mathbf{B}^{-1}\mathbf{A}$.

A.8   The Determinant of a Matrix

The determinant of a matrix is defined as a function that assigns a real-valued number to this matrix. Determinants are defined only for square matrices. Therefore, we consider only square matrices in this section. Rather than going into technical details concerning the calculation of determinants we list five characteristics of determinants:

1. A matrix $\mathbf{A}$ is nonsingular if and only if the determinant, abbreviated $|\mathbf{A}|$ or $det(\mathbf{A})$, is different than 0.

2. If a matrix $\mathbf{A}$ is nonsingular, the following holds:

$$det(\mathbf{A}^{-1}) = \frac{1}{det(\mathbf{A})}.$$

3. The determinant of a diagonal or triangular matrix equals the product of its diagonal elements. Let **A** be, for instance,

$$\mathbf{A} = \begin{pmatrix} a_{11} & a_{12} & \cdots & a_{1n} \\ 0 & a_{22} & \cdots & a_{2n} \\ \vdots & \vdots & \ddots & \vdots \\ 0 & 0 & \cdots & a_{nn} \end{pmatrix},$$

then

$$det(\mathbf{A}) = a_{11}a_{22}\cdots a_{nn}.$$

4. The determinant of a product of two matrices equals the product of the two determinants,

$$det(\mathbf{AB}) = det(\mathbf{A})det(\mathbf{B}).$$

5. The determinant of the transpose of a matrix equals the determinant of the original matrix,

$$det(\mathbf{A'}) = det(\mathbf{A}).$$

A.9 Rules for Operations with Matrices

In this section we present a selection of operations that can be performed with vectors and matrices. We do not provide examples or proofs. Readers are invited to create examples for each of these operations. Operations include the simpler operations of addition and subtraction as well as determinants and the more complex operations of multiplication and inversion. It is tacitly assumed that all matrices $\mathbf{A}, \mathbf{B}, \ldots$, have admissible dimensions for the corresponding operations.

1. When adding matrices the order is unimportant. One says that the addition of matrices is commutative:

$$\mathbf{A} + \mathbf{B} = \mathbf{B} + \mathbf{A}.$$

2. Here is a simple extension of the previous rule:

$$(\mathbf{A} + \mathbf{B}) + \mathbf{C} = \mathbf{A} + (\mathbf{B} + \mathbf{C}).$$

3. When multiplying three matrices the result does not depend on which two matrices are multiplied first as long as the order of multiplication is preserved:

$$(\mathbf{AB})\mathbf{C} = \mathbf{A}(\mathbf{BC}).$$

4. Factoring out a product term in matrix addition can be performed just as in ordinary algebra:

$$\mathbf{C}(\mathbf{A} + \mathbf{B}) = \mathbf{CA} + \mathbf{CB}.$$

5. Here is an example of factoring out a scalar as product term:

$$k(\mathbf{A} + \mathbf{B}) = k\mathbf{A} + k\mathbf{B} \ , \ \text{where } k \text{ is a scalar}$$

6. Multiplying first and then transposing is equal to transposing first and then multiplying, but in reversed order:

$$(\mathbf{AB})' = \mathbf{B}'\mathbf{A}'.$$

7. The transpose of a transposed matrix equals the original matrix:

$$(\mathbf{A}')' = \mathbf{A}.$$

8. Adding first and then transposing is equal to transposing first and then adding:

$$(\mathbf{A} + \mathbf{B})' = \mathbf{A}' + \mathbf{B}'.$$

9. Multiplying first and then inverting is equal to inverting first and then multiplying, but in reversed order:

$$(\mathbf{AB})^{-1} = \mathbf{B}^{-1}\mathbf{A}^{-1}.$$

10. The inverse of an inverted matrix equals the original matrix:

$$(\mathbf{A}^{-1})^{-1}.$$

11. Transposing first and then inverting is equal to inverting first and then transposing:

$$(\mathbf{A}')^{-1} = (\mathbf{A}^{-1})'$$

A.10 Exercises

1. Consider the following three matrices, $\mathbf{A}$, $\mathbf{B}$, and $\mathbf{C}$. Create the transpose for each of these matrices.

$$\mathbf{A} = \left(\begin{array}{ccc} 1 & 3 & -9 \\ 9 & 7 & -2 \end{array} \right), \ \mathbf{B} = \left(\begin{array}{cc} -1 & -1 \\ -7 & 15 \\ 99 & -8 \end{array} \right), \ \mathbf{C} = \left(\begin{array}{cc} 34 & 43 \\ -6 & 65 \end{array} \right).$$

2. Consider, again, the above matrices, $\mathbf{A}$, $\mathbf{B}$, and $\mathbf{C}$. Transpose $\mathbf{B}$ and add it to $\mathbf{A}$. Transpose $\mathbf{A}$ and add it to $\mathbf{B}$. Transpose one of the resulting matrices and compare it to the other result of addition. Perform the same steps with subtraction. Explain why $\mathbf{A}$ and $\mathbf{B}$ cannot be added to each other as they are. Explain why $\mathbf{C}$ cannot be added to either $\mathbf{A}$ or $\mathbf{B}$, even if it is transposed.

3. Consider the matrix, $\mathbf{A}$, and the vector, $\mathbf{v}$, below:

$$\mathbf{A} = \left(\begin{array}{cc} -2 & 4 \\ 15 & 1 \\ -1 & 3 \end{array} \right), \ \mathbf{v} = \left(\begin{array}{c} 12 \\ -2 \end{array} \right).$$

Postmultiply $\mathbf{A}$ by $\mathbf{v}$.

4. Consider the matrix, $\mathbf{X}$, below

$$\mathbf{X} = \left(\begin{array}{ccc} 2 & 4 & 17 \\ 1 & 2 & -3 \\ 9 & 2 & -9 \end{array} \right).$$

Multiply the matrix, $\mathbf{X}$, by the scalar -3.

5. Consider the matrices, $\mathbf{A}$ and $\mathbf{B}$, below

$$\mathbf{A} = \begin{pmatrix} 2 & 3 \\ 4 & 3 \end{pmatrix}, \ \mathbf{B} = \begin{pmatrix} 2 & -2 & 4 & 12 \\ 1 & 32 & 6 & -1 \end{pmatrix}.$$

Multiply these matrices with each other. (Hint: postmultiply $\mathbf{A}$ with $\mathbf{B}$.)

6. Consider the two vectors, $\mathbf{v}$ and $\mathbf{w}$, below:

$$\mathbf{v}' = \begin{pmatrix} 2 & 4 & 1 & 3 \end{pmatrix}, \ \mathbf{w}' = \begin{pmatrix} 8 & 2 & -3 & 1 \end{pmatrix}.$$

Perform the necessary operations to create (1) the inner product and (2) the outer product of these two vectors. Add these two vectors to each other. Subtract $\mathbf{w}$ from $\mathbf{v}$.

7. Consider the diagonal matrix $\mathbf{D} = diag[34, 43, -12, 17, 2]$. Calculate the inverse of this matrix. Multiply $\mathbf{D}$ with $\mathbf{D}^{-1}$. Discuss the result.

8. Consider the following matrix:

$$\mathbf{A} = \begin{pmatrix} 4 & 3 \\ 3 & 3 \end{pmatrix}$$

Create the inverse of this matrix.

9. Find the 2*3 matrix whose elements are its cell indices.

Appendix B

BASICS OF DIFFERENTIATION

This excursus reviews basics of differentiation. Consider function $f(x)$. Let x be a real number and $f(x)$ be a real-valued function. For this function we can determine the slope of the tangent, that is the function, that touches $f(x)$ at a given point, x_0. In Figure B.1 the horizontal straight line is the tangent for the smallest value of $f(x)$. The arrow indicates where the tangent touches the square function.

The tangent of a curve at a specific point x_0 is determined by knowing the slope of the curve at that point. The slope at x_0 can be obtained by inserting x_0 into the *first derivative*,

$$\frac{d}{dx}f(x),$$

of the function $f(x)$. In other words,

$$\frac{d}{dx}f(x_0)$$

yields the slope of the curve at x_0 and thus the slope of the tangent of $f(x)$ at x_0. Often the first derivative is denoted by $f'(x)$. At an extremum the slope of $f(x)$ is zero. Therefore, differentiation is a tool for determining locations where a possible extremum might occur. This is done by setting

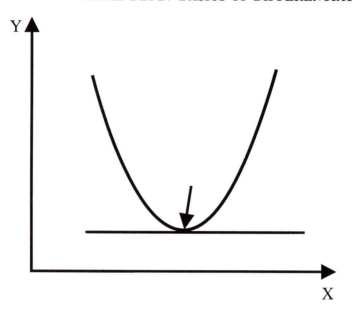

Figure B.1: Minimum of square curve.

the first derivative of $f(x)$ to zero and solving for x. This yields the points where an extremum is possible. Therefore, knowledge of how derivatives are created is crucial for an understanding of function minimization. We now review six rules of differentiation.

1. The first derivative of a constant function, $f(x) = k$, is zero:

$$\frac{d}{dx}k = 0.$$

2. The first derivative of the function $f(x) = x$ is 1:

$$\frac{d}{dx}x = 1.$$

3. Rule: Creating the first derivative of a function, $f(x)$, that has the form $f(x) = x^n$ proceeds in two steps:

   - subtract 1 from the exponent

- place the original exponent as a factor of x in the function

Applying these two steps yields as the derivative of x^n

$$\frac{d}{dx}x^n = nx^{n-1}.$$

Notice that Rule 2 can be derived from Rule 3. We obtain

$$\frac{d}{dx}x^1 = 1x^0 = 1.$$

4. If we want to differentiate a function that can be written as $kf(x)$, where k is a constant, then we can write this derivative in general form as

$$\frac{d}{dx}kf(x) = k\frac{d}{dx}f(x).$$

5. The first derivative of the sum of two functions, $f_1(x)$ and $f_2(x)$, equals the sum of the first derivatives of the two functions,

$$\frac{d}{dx}(f_1(x) + f_2(x)) = \frac{d}{dx}f_1(x) + \frac{d}{dx}f_2(x),$$

or, more generally,

$$\frac{d}{dx}\sum_{i=1}^{n}f_i(x) = \sum_{i=1}^{n}\frac{d}{dx}f_i(x).$$

In words, summation and differentiation can be interchanged.

6. The first derivative of a function $f(y)$ which takes as its argument another function $g(x) = y$ is given by the *chain rule*:

$$\frac{d}{dx}f(g(x)) = \frac{d}{dy}f(y)\frac{d}{dx}g(x).$$

For example, let $f(y) = y^2$ and $y = g(x) = (2 - x)$, that is, $f(x) = (2 - x)^2$. Now, the chain rule says that we should multiply the derivative of f with respect to y, which in the example is $2y$, with the derivative of g with respect to x, which is -1. We obtain

$$\frac{d}{dx}f(g(x)) = 2y(-1) = 2(2-x)(-1) = 2x - 4.$$

after substituting $(2-x)$ for y.

So far we have only considered functions that take a single argument, usually denoted as x, and return a real number denoted as $f(x)$. To obtain the least squares estimates we have to be able to handle functions that, while still returning a real number, take two arguments, that is, $f(x,y)$. These functions are defined over the $x-y$ plane and represent a surface in three-dimensional space. Typically, these functions are differentiable as well. We will not try to explain differential calculus for real-valued functions in n-space. All we need is the notion of partial derivation, and this is quite easy. We merely look at $f(x,y)$ as if it were only a function of x, treating y as a constant, and vice versa. In other words, we pretend that $f(x,y)$ is a function taking a single argument. Therefore, all the above rules for differentiation apply. Just the notation changes slightly; the partial derivative is denoted as

$$\frac{\partial}{\partial x}f(x,y)$$

if we treat y as a constant and as

$$\frac{\partial}{\partial y}f(x,y)$$

if x is treated as a constant. Again, what we are after is finding extrema. A result from calculus states that a necessary condition for an extremum at point (x_0, y_0) in the xy plane is that the two partial derivatives are zero at that point, that is,

$$\frac{\partial}{\partial x}f(x_0, y_0) = 0 \quad \text{and} \quad \frac{\partial}{\partial y}f(x_0, y_0) = 0.$$

Appendix C

BASICS OF VECTOR DIFFERENTIATION

Consider the inner product $\mathbf{a}'\mathbf{x}$, where $\mathbf{a}$ and $\mathbf{x}$ are vectors each having p elements. $\mathbf{a}$ is considered a vector of constants and $\mathbf{x}$ is a vector containing p variables. We can write $f(x_1, x_2, \ldots, x_p) = \mathbf{a}'\mathbf{x}$, thus considering the inner product $\mathbf{a}'\mathbf{x}$ a real valued function of the p variables. Now, the partial derivatives of f with respect to each x_j are easily obtained as

$$\frac{\partial}{\partial x_j}\mathbf{a}'\mathbf{x} = \frac{\partial}{\partial x_j}\sum_{i=1}^{p} a_i x_i = \sum_{i=1}^{p} \frac{\partial}{\partial x_j} a_i x_i = a_j.$$

In words this says that each partial derivative of $\mathbf{a}'\mathbf{x}$ with respect to x_j just yields the corresponding constant a_j.

We now define the derivative of f with respect to the vector $\mathbf{x}$ as

$$\frac{\partial}{\partial \mathbf{x}} f(\mathbf{x}) = \begin{pmatrix} \frac{\partial}{\partial x_1} f(\mathbf{x}) \\ \frac{\partial}{\partial x_2} f(\mathbf{x}) \\ \vdots \\ \frac{\partial}{\partial x_p} f(\mathbf{x}) \end{pmatrix},$$

and for the special case that $f(\mathbf{x}) = \mathbf{a}'\mathbf{x}$ we obtain as the derivative with respect to $\mathbf{x}$

$$\frac{\partial}{\partial \mathbf{x}} \mathbf{a}' \mathbf{x} = \begin{pmatrix} a_1 \\ a_2 \\ \vdots \\ a_p \end{pmatrix} = \mathbf{a},$$

that is, we obtain the vector $\mathbf{a}$.

We formulate this as Rule 1.

$$\text{Rule 1:} \quad \frac{\partial}{\partial \mathbf{x}} \mathbf{a}' \mathbf{x} = \mathbf{a}.$$

The following two rules follow accordingly.

$$\text{Rule 2:} \quad \frac{\partial}{\partial \mathbf{x}} \mathbf{x}' \mathbf{a} = \mathbf{a}.$$

$$\text{Rule 3:} \quad \frac{\partial}{\partial \mathbf{x}} \mathbf{a} = \mathbf{0},$$

where $\mathbf{0}$ denotes a vector of zeros and $\mathbf{a}$ is a vector or a real-valued number. In both cases the result will be the vector $\mathbf{0}$. In words Rule 3 says that each partial derivative of a constant vector or a constant is zero.

Combining all these partial derivatives into a vector yields $\mathbf{0}$. It should be noted that these rules are very similar to the corresponding rules for differentiation for real-valued functions that take a single argument. This holds as well for the following:

$$\text{Rule 4:} \quad \frac{\partial}{\partial \mathbf{x}} \mathbf{x}' \mathbf{x} = 2\mathbf{x}.$$

Combining the partial derivative of the inner product of a vector with itself yields the same vector with each element doubled.

$$\text{Rule 5:} \quad \frac{\partial}{\partial \mathbf{x}} \mathbf{x}' \mathbf{A} \mathbf{x} = (\mathbf{A}\mathbf{A}')\mathbf{x}.$$

If $\mathbf{A}$ is symmetric, functions of the form $\mathbf{x}'\mathbf{A}\mathbf{x}$ are called quadratic forms. In this case, $\mathbf{A} = \mathbf{A}'$ and Rule 5 simplifies to $\frac{\partial}{\partial \mathbf{x}} \mathbf{x}' \mathbf{A} \mathbf{x} = 2\mathbf{A}\mathbf{x}$.

With theses rules we are in the position to derive the general OLS solution very easily. For further details on vector differentiation, see for instance, Mardia, Kent, and Bibby (1979), Morrison (1990).

Appendix D

POLYNOMIALS

The following excursus presents a brief introduction to polynomials and systems of polynomials (see, for instance, Abramowitzm & Stegun, 1972). Readers familiar with these topics can skip the excursus.

Functions that

1. are defined for all real-valued x and

2. do not involve any divisions

are called *polynomials*of the form given in (D.1)

$$y = b_0 + b_1 x + b_2 x^2 + \cdots = \sum_{j=0}^{J} b_j x^j, \qquad (D.1)$$

where the real-valued b_j are the polynomial parameters. The vectors **x** contain x values, termed polynomial coefficients. These coefficients describe polynomials given in standard form.

It is important to note that the formula given in (D.1) is linear in its parameters. The x values are raised to powers of two and greater. Therefore, using polynomials to fit data can still be accomplished within the framework of the General Linear Model.

The highest power to which an x value is raised determines the degree of the polynomial, also termed the order of the polynomial. For example, if the highest power to which an x value is raised is 4, the polynomial is called a fourth-degree polynomial or fourth-order polynomial. If the highest power is J, the polynomial is a Jth-order polynomial.

Figure 7.2 presents four examples of polynomials – one first-, one second-, one third-, and one fourth-order polynomial, all for seven observation points.

The polynomial coefficients used to create Figure 7.2 appear in Table D.1. Textbooks of analysis of variance contain tables with polynomial coefficients that typically cover polynomials up to fifth order and up to 10 values of X; see Fisher and Yates (1963, Table 23) or Kirk (1995, Table E10).

Table D.1: *Polynomial Coefficients for First-, Second-, Third-, and Fourth-Order Polynomials for Seven Values of X*

Predictor	Polynomials (Order)			
Values	First	Second	Third	Fourth
1	−3	5	−1	3
2	−2	0	1	−7
3	−1	−3	1	1
4	0	−4	0	6
5	1	−3	−1	1
6	2	0	−1	−7
7	3	5	1	3

The straight line in Figure 7.2 displays the coefficients in the column for the first order polynomial in Table D.1. The graph of the second order polynomial is U-shaped. Quadratic polynomials have two parameters that can be interpreted. Specifically, these are the slope parameter from the linear polynomial and the parameter for the curvature.[1] The sign of this parameter indicates what type of extremum the curve has. A positive

---

[1] In the following sections we apply the Hierarchy Principle. This principle requires that a polynomial of degree J contains all lower order terms, that is, the $J − 1th$ term, the $J − 2nd$ term, ..., and the $J − Jst$ term (the constant). The reason for adhering to the Hierarchy Principle is that it is rarely justifiable to only use a specific polynomial term. For instance, in simple regression, one practically always uses both b_0 and b_1. Omitting b_0 forces the intercept to be zero which is reasonable only if the dependent variable, Y, is centered (or when regression analysis is based on a correlation matrix).

parameter indicates that the curve is U-shaped, that is, has a minimum. A negative parameter indicates that the curve has a maximum. This curve is inversely U-shaped. The size of the parameter indicates how flat the curve is. Small parameters suggest flat curves, large parameters suggest tight curves.

First- and second-order polynomials do not have inflection points, that is, they do not change direction. "Changing direction" means that the curve changes from a curve to the right to a curve to the left and vice versa. A look at Figure 7.2 suggests that neither the first-order polynomial nor the second-order polynomial change direction in this sense. In contrast, the third and the quartic polynomials do have inflection points. For example, the line of the third-order polynomial changes direction at $X = 4$ and $Y = 0$.

The size of the parameter of the third-order polynomial indicates how tight the curves of the polynomial are. Large parameters correspond with tighter curves. Positive parameters indicate that the last "arm" of the polynomial goes upward. This is the case for all four polynomials in Figure 7.2. Negative parameters indicate that the last "arm" goes downward.

The fourth-order polynomial has two inflection points. The curve in Figure 7.2 has its two inflection points at $X = 2.8$ and $X = 5.2$, both at $Y = 0$. The magnitude of the parameter of fourth-order polynomials indicates, as for the other polynomials of second and higher order, how tight the curves are. The sign of the parameter indicates the direction of the last arm. with positive signs corresponding to an upward direction of the last arm.

D.1 Systems of Orthogonal Polynomials

This section introduces readers to a special group of polynomials, orthogonal polynomials. These polynomials are equivalent to other types of polynomials in many important characteristics; however, they have unique characteristics that make them particularly useful for regression analysis. Most importantly, orthogonal polynomials are "independent" of each other. Thus, polynomials of different degrees can be added or eliminated without affecting the magnitude of the parameters already in the equation.

Consider a researcher that has estimated a third-order polynomial to smooth the data collected in an experiment on the Yerkes-Dodson-Law (see Chapter 7). After depicting the data, the researcher realizes that a third-order polynomial may not be necessary, and a quadratic curve may be sufficient. If the researcher has fit an orthogonal third-order polynomial, all that needs to be done to switch to a second-order polynomial is to drop the third-order term. Parameters for the lower order terms will remain unaffected.

To introduce systems of orthogonal polynomials, consider the expression

$$y = b_0\xi'_0 + b_1\xi'_1 + b_2\xi'_2 + \cdots , \tag{D.2}$$

where the ξ'_i, with $i = 0, 1, \ldots$, denote polynomials of ith order. Formula (D.2) describes a system of polynomials. In order to describe a system of orthogonal polynomials, any two different polynomials on the system must fulfill the orthogonality condition

$$\sum_i y_{ij}y_{ki} = 0, \quad \text{for } j \neq k,$$

where j and k index polynomials, and i indexes cases, for instance, subjects. This condition requires the inner product of any two different vectors of polynomial coefficients to be zero. Consider, for example, the orthogonal polynomial coefficients in Table D.1. The inner product of the coefficients for the third- and the fourth-order polynomials is $(-1)*3+1*(-7)+1*1+0*6+(-1)*1+(-1)*(-7)+1*3 = 0$. Readers are invited to vector-wise sum polynomial coefficients and to calculate the other inner products for the vectors in Table D.1 and to decide whether the vectors in this table stem from a system of orthogonal polynomials. It should be noted that this condition cannot always be met by centering.

While not part of the orthogonality condition, the following condition is also often placed, for scaling purposes:

$$\sum_i y_i = 0.$$

In words, this condition requires that the sum of y values, that is, the sum of polynomial coefficients, equals zero. This can be obtained for any array of real-valued measures by, for example, centering them. Consider the following example. The array $\mathbf{y'} = (1, 2, 3)$ is centered to be $\mathbf{y'_c} = (-1, 0, 1)$. Whereas the sum of the components, $\mathbf{y}$, equals 6, the sum of the components of $\mathbf{Y_c}$ equals 0.

Equation (D.2) contains the ξ'_i, for $i \geq 0$, as placeholders for polynomials. The following formula provides an explicit example of a system of two polynomials, one first- and one second-order polynomial:

$$y = \beta_1(\alpha_{10} + \alpha_{11}x) + \beta_2(\alpha_{20} + \alpha_{21}x + \alpha_{22}x^2).$$

In this equation, the β_j, $for j = 1, 2$, are the parameters (weights) for the polynomials, and the α_{il}, $for j = 1, 2, and l = 0, 1, 2$, are the parameters within the polynomials.

As we indicated before, the parameter estimates for orthogonal polynomials are independent of each other. Most importantly, the β_j remain unchanged when one adds or eliminates any other parameter estimate for a polynomial of different degree.

D.2 Smoothing Series of Measures

Polynomials can be used to smooth any series of measures. If there are J measures, the polynomial of degree $J-1$ will always go exactly through all measures. For reasons of scientific parsimony, researchers typically strive for polynomials of degree lower than $J - 1$. The polynomial of degree $J - 1$ has J parameters, that is, as many parameters as there are data points. Thus, there is no data reduction. The polynomial of degree $J - 2$ has $J - 1$ parameters. It may not go exactly through all data points. However, considering the ubiquitous measurement error, one may be in a position where the polynomial describes the series very well nevertheless.

However, there are some problems with polynomial approximation. Two of these problems will be briefly reviewed here. First, while one can approximate any series of measures as closely as needed, this may not always be possible using polynomials of low order. This applies in particular to cyclical processes such as seasonal changes, rhythms, and circadian changes. It also applies to processes that approach an asymptote or taper off. Examples of the former include learning curves, and examples of the latter include forgetting curves. Thus, one may approximate these types of processes using functions other than polynomials.

Second, as is obvious from Figure 7.2, polynomials assume increasingly extreme values when x increases (or decreases). Thus, using polynomial functions for purposes of extrapolation only makes sense if one assumes that the process under study also assumes these extreme values. This applies accordingly to interpolation (see the sections on extrapolation and interpolation in Section 2.4).

Appendix E

DATA SETS

E.1 Recall Performance Data

This is the data set used to illustrate the variable selection techniques.

AGE	EG1	SEX	HEALTH	READ	EDUC	CC1	CC2	OVC	TG	REC
19	2	1	1	3	4	14	56	0.24	1	120
19	2	1	1	4	4	36	36	0.15	1	126
18	2	1	2	4	4	8	33	0.65	1	121
20	2	1	1	4	4	24	38	0.21	1	90
21	2	1	1	3	4	7	33	0.70	1	94
20	2	1	1	2	4	15	37	0.29	1	98
18	2	1	1	3	4	12	38	0.42	2	94
19	2	1	2	3	4	14	50	0.27	2	53
19	2	2	2	2	4	13	42	0.35	2	79
19	2	1	2	3	4	14	48	0.28	2	72
19	2	1	1	4	4	10	35	0.54	2	63
21	2	2	3	4	4	7	30	0.90	2	82
23	4	2	1	4	6	11	62	0.28	1	91
30	4	1	1	4	6	11	48	0.36	1	72
24	4	1	1	4	6	5	36	1.00	1	80
23	4	2	1	4	5	11	31	0.56	1	63
28	4	2	2	4	6	10	46	0.41	1	69
27	4	1	2	3	6	10	36	0.49	1	63
30	4	1	2	3	5	9	56	0.38	1	53

25	4	1	1	4	6	14	28	0.49	1	40
29	4	2	1	4	6	7	24	1.00	1	38
22	4	1	1	4	5	7	24	1.00	1	38
28	4	1	1	4	6	6	37	0.73	1	44
22	4	1	2	2	4	5	40	0.95	1	114
21	4	2	2	2	4	7	23	0.94	1	21
18	4	2	1	3	4	7	43	0.60	1	80
18	4	1	2	2	4	16	37	0.32	1	71
21	4	1	2	2	4	8	31	0.52	1	49
25	4	2	1	4	5	11	38	0.43	1	41
20	4	2	1	4	4	7	39	0.67	1	74
23	4	1	3	4	5	5	28	0.93	1	43
20	4	2	1	2	4	5	36	0.78	1	61
21	4	1	1	4	4	7	40	0.64	1	89
20	4	2	1	2	4	8	35	0.57	1	54
22	4	1	1	2	4	13	39	0.35	1	64
21	4	2	2	4	5	6	48	0.62	1	88
21	4	1	1	2	4	6	33	0.90	1	44
20	4	2	1	3	4	18	32	0.33	1	26
20	4	1	2	3	4	7	43	0.60	1	27
25	4	1	2	4	4	6	27	1.00	1	48
20	4	1	1	4	4	11	35	0.47	1	48
21	4	1	1	3	4	7	36	0.76	1	61
23	4	2	1	2	4	12	34	0.46	1	64
21	4	1	1	2	4	9	38	0.50	1	33
20	4	1	2	3	4	6	48	0.38	1	80
22	4	1	1	4	5	18	33	0.32	1	97
20	4	1	2	3	4	12	40	0.40	2	58
21	4	1	1	3	4	6	28	0.76	2	30
24	4	1	1	3	5	8	43	0.53	2	43
20	4	1	1	2	4	8	35	0.65	2	77
26	4	1	1	4	5	5	28	1.00	2	26
25	4	1	1	4	6	7	44	0.61	2	46
22	4	1	2	4	5	11	57	0.30	2	38
28	4	1	1	4	6	13	41	0.36	2	49
26	4	2	1	4	5	11	52	0.33	2	42
29	4	1	1	4	4	10	26	0.73	2	51
20	4	1	2	2	4	9	39	0.46	2	35
18	4	1	1	2	4	8	38	0.60	2	18
21	4	2	1	3	4	8	36	0.55	2	31

25	4	1	2	2	4	8	35	0.61	2	38
21	4	1	2	4	4	10	55	0.35	2	19
18	4	1	2	2	4	8	39	0.52	2	12
25	4	2	1	3	5	16	69	0.17	2	86
18	4	1	1	2	4	6	33	0.90	2	55
18	4	1	1	2	4	5	72	0.53	2	38
20	4	1	1	2	4	9	40	0.42	2	39
21	4	2	1	3	4	6	45	0.59	2	39
20	4	2	2	4	4	11	45	0.39	2	47
22	4	1	1	4	5	16	48	0.25	2	30
21	4	2	1	3	4	4	28	1.00	2	20
19	4	1	1	4	4	6	38	0.65	2	35
20	4	2	1	4	4	18	43	0.25	2	53
20	4	1	1	2	4	8	31	0.76	2	20
18	4	1	2	2	4	12	35	0.40	2	51
20	4	1	2	4	4	9	24	0.77	2	17
20	4	1	2	3	4	6	32	1.00	2	51
20	4	1	1	3	4	9	42	0.50	2	51
46	2	2	1	4	6	13	35	0.35	1	145
37	2	1	2	4	4	10	35	0.54	1	99
47	2	1	1	4	6	24	24	0.32	1	79
34	2	1	2	3	4	13	35	0.42	1	121
37	2	1	1	5	6	9	53	0.40	1	165
45	2	2	1	4	6	12	40	0.40	1	66
40	2	1	2	5	5	7	39	0.52	2	28
31	2	2	1	5	6	15	46	0.28	2	78
40	2	2	2	1	4	6	37	0.68	2	34
47	2	1	2	1	4	10	20	0.95	2	17
48	2	1	1	3	4	5	30	1.00	2	23
38	2	1	1	3	4	14	32	0.42	2	16
47	2	1	2	4	4	6	38	0.78	2	118
48	4	1	2	4	4	7	33	0.74	1	60
47	4	1	1	3	4	12	39	0.38	1	67
46	4	1	1	1	4	8	36	0.48	1	40
44	4	1	1	3	5	12	45	0.35	1	72
46	4	1	1	2	4	11	29	0.60	1	55
46	4	1	1	4	6	20	38	0.25	1	48
44	4	2	2	2	5	7	40	0.68	1	107
46	4	1	3	4	7	6	32	0.78	1	61
49	4	1	3	3	4	24	31	0.25	1	60

40	4	1	1	4	5	14	27	0.47	1	53
50	4	1	1	4	5	29	54	0.12	1	76
48	4	1	1	2	4	10	32	0.56	1	89
43	4	1	1	3	6	12	39	0.40	1	84
47	4	2	1	4	7	28	40	0.17	1	80
44	4	1	1	4	4	12	32	0.47	1	44
43	4	1	1	3	4	15	39	0.32	1	69
40	4	1	1	4	5	14	41	0.33	1	79
44	4	2	2	4	6	17	78	0.14	1	50
40	4	1	1	1	6	22	54	0.16	1	46
40	4	1	1	4	6	24	38	0.23	1	72
49	4	2	3	3	3	6	42	0.68	2	17
45	4	2	1	4	6	8	34	0.63	2	54
47	4	1	1	2	5	21	34	0.27	2	40
39	4	1	3	3	4	12	26	0.61	2	34
38	4	1	2	3	4	6	27	1.00	2	33
38	4	1	2	4	4	16	30	0.39	2	42
44	4	2	1	4	5	10	41	0.46	2	22
49	4	1	2	4	5	27	52	0.13	2	28
51	4	1	2	3	5	11	37	0.46	2	43
42	4	1	1	4	6	23	47	0.17	2	64
42	4	2	1	3	6	24	40	0.20	2	38
42	4	1	1	2	5	7	37	0.69	2	27
40	4	1	1	4	6	10	40	0.45	2	73
41	4	1	1	3	4	8	33	0.65	2	33
46	4	1	3	3	4	10	48	0.37	2	38
40	4	2	2	2	6	15	38	0.33	2	43
42	4	1	2	4	6	17	45	0.22	2	39
47	4	1	1	4	4	9	41	0.51	2	49
41	4	2	1	4	5	26	32	0.23	2	69
48	4	1	2	3	4	11	34	0.46	2	31
46	4	1	1	4	5	10	40	0.41	2	54
49	4	2	3	4	4	8	60	0.39	2	18
67	2	2	2	3	5	5	34	1.00	1	87
61	2	1	2	5	5	4	40	1.00	1	97
65	2	1	2	5	4	7	37	0.65	1	59
61	2	1	2	4	4	7	44	0.61	1	130
70	2	2	1	4	6	13	39	0.34	1	111
68	2	1	1	5	5	8	55	0.43	1	73
69	2	1	2	3	4	6	25	1.00	2	38

64	2	2	2	4	5	8	36	0.62	2	42
67	2	1	2	4	4	6	32	1.00	2	49
67	2	1	2	5	5	8	26	0.90	2	67
68	2	1	3	4	4	7	24	1.00	2	50
69	4	1	1	4	5	7	49	0.56	1	47
60	4	1	2	4	5	8	30	0.63	1	52
64	4	1	1	3	6	7	31	0.86	1	37
63	4	1	2	4	5	12	42	0.38	1	67
64	4	1	1	3	5	8	34	0.63	1	67
63	4	2	1	4	6	15	46	0.28	1	65
60	4	2	2	4	7	16	24	0.33	1	50
60	4	1	1	3	6	9	43	0.46	1	57
61	4	1	2	3	4	17	33	0.34	1	93
66	4	2	1	2	4	4	30	1.00	1	37
61	4	1	1	2	5	21	30	0.30	1	58
67	4	2	1	3	7	14	38	0.36	1	57
59	4	1	1	3	5	17	32	0.35	1	40
69	4	1	2	3	6	8	45	0.52	1	37
73	4	2	3	2	4	6	35	0.76	1	14
60	4	1	1	3	5	6	42	0.72	1	11
60	4	2	2	3	4	11	49	0.33	1	51
60	4	1	2	4	5	20	42	0.23	1	54
67	4	1	2	3	6	17	35	0.30	1	84
63	4	2	1	3	7	21	45	0.20	1	25
66	4	2	1	3	4	5	28	1.00	2	16
63	4	2	2	4	5	6	27	1.00	2	20
59	4	1	1	2	4	8	29	0.78	2	42
70	4	1	2	3	7	9	31	0.67	2	39
61	4	1	2	4	5	7	27	0.95	2	29
61	4	1	1	2	4	6	33	0.90	2	21
64	4	1	1	3	4	11	36	0.45	2	41
67	4	2	2	3	5	5	38	0.68	2	29
64	4	1	1	4	6	17	33	0.34	2	45
67	4	1	1	4	7	7	77	0.35	2	50
63	4	1	2	4	5	12	34	0.46	2	22
60	4	2	1	4	6	11	26	0.66	2	25
59	4	2	1	2	7	22	42	0.21	2	34
62	4	1	2	2	4	6	27	0.94	2	29
66	4	1	1	4	4	6	42	0.52	2	45
63	4	1	1	2	5	12	32	0.50	2	34

67	4	2	2	3	7	18	43	0.23	2	47
59	4	2	2	3	5	5	30	1.00	2	110
69	4	1	2	4	4	5	42	0.90	2	26
65	4	2	1	4	6	12	42	0.34	2	42

E.2 Examination and State Anxiety Data

This is the data set that was used to illustrate robust modeling of longitudinal data.

```
          Females           Males
          t1 t2 t3 t4 t5  t1 t2 t3 t4 t5

          13 17 18 20 24   6 14 22 20 24
          26 31 33 38 42   4 11 14 12 23
          13 17 24 29 32  17 25 26 29 38
          22 24 26 27 29  19 22 26 30 34
          18 19 19 22 30  12 21 21 23 24
          32 31 30 31 32  11 16 20 19 22
          16 16 21 27 30  14 23 26 29 33
          18 22 25 29 35   9 18 20 20 24
          14 17 23 21 25  12 16 23 26 32
          20 19 23 25 28  11 13 17 14 20
          17 24 24 24 26  12 20 22 23 31
          22 22 24 24 26   7 15 16 18 26
          19 24 27 29 34   2 10 16 16 22
          30 29 29 28 28   3 10 13 14 24
          21 25 24 28 30   3 13 13 15 23
          16 20 19 21 24   9 17 21 22 25
          24 23 27 28 33  16 23 23 29 35
          19 20 22 25 30  14 24 23 24 30
          21 27 30 30 30   6 15 17 18 26
          17 17 20 23 28   6 16 16 15 25
          15 20 23 21 28   1  6 11 15 16
          20 22 26 27 31  14 16 21 21 30
          22 22 24 25 26   6 15 16 17 24
          23 26 28 30 32  13 20 27 27 32
          10 13 14 16 23   5 17 21 19 22
```

```
28 29 27 30 32  8 17 21 23 28
25 27 28 29 31  8 15 18 19 26
26 28 30 31 35 15 23 24 27 28
23 25 27 27 30 16 23 23 24 29
25 32 34 33 38 12 21 26 27 35
23 32 34 35 32 15 23 25 24 30
20 19 19 23 25  5 10 16 21 24
20 24 30 33 38 15 19 26 25 30
16 22 23 26 32  3 11 12 14 19
24 29 29 33 37 11 20 21 19 22
17 20 23 26 32  7 15 18 21 31
11 13 17 22 29 20 27 32 30 37
25 26 27 28 34  4  8 11 13 22
19 21 27 26 27  3 11 17 20 25
22 25 28 32 38 18 24 28 26 29
21 24 23 23 23  1  8 10 12 16
15 17 16 17 16  5 11 15 21 29
23 21 16 18 21 15 23 23 27 28
21 24 24 28 30 11 15 20 18 27
17 22 24 22 26 11 21 25 23 27
31 33 36 38 42  0  7 10 14 20
21 24 25 28 33 15 19 25 26 30
19 26 27 27 30  8 14 15 17 25
22 28 26 27 28 13 21 26 29 35
20 22 26 30 30  6 19 21 25 30
```

References

Abramowitzm, M., & Stegun, I. (1972). *Handbook of mathematical functions.* New York: Dover.

Afifi, A. A., & Clark, V. (1990). *Computer-aided multivariate analysis* (2nd ed.). New York: von Nostrand Reinhold.

Aiken, L. S., & West, S. G. (1991). *Multiple regression: Testing and interpreting interactions.* Newbury Park: Sage.

Alexander, R. A., & Govern, D. M. (1994). A new and simpler approximation for ANOVA under variance heterogeneity. *Journal of Educational Statistics, 1*, 91–101.

Arnold, S. F. (1990). *Mathematical statistics.* Englewood Cliffs, New Jersey: Prentice-Hall.

Atkinson, A. C. (1986). Comment: Aspects of diagnostic regression analysis. *Statistical Science, 1*, 397–402.

Ayres, F. J. (1962). *Theory and problems of matrices.* New York: McGraw-Hill.

Box, G. E. P., & Cox, D. R. (1964). An analysis of transformations (with discussion). *Journal of the Royal Statistical Society, B, 26*, 211–246.

Bryk, A. S., & Raudenbush, S. W. (1992). *Hierarchical linear models.* Newbury Park, CA: Sage.

Christensen, R. (1996). *Plane answers to complex questions* (2nd ed.). New York: Springer-Verlag.

Cohen, J. (1978). Partialed products are interactions; partialed powers are curved components. *Psychological Bulletin, 85*, 856–866.

Cohen, J. (1988). *Statistical power analysis for the behavioral sciences* (2nd ed.). Hillsdale, NJ: Lawrence Erlbaum Associates.

Cox, D. R. (1958). *Planning of experiments.* New York: John Wiley & Sons.

Cronbach, L. J. (1987). Statistical tests for moderator variables: Flaws in analysis recently proposed. *Psychological Bulletin, 102,* 414–417.

Darlington, R. B. (1990). *Regression and linear models.* New York: McGraw-Hill.

de Shon, R. P., & Alexander, R. A. (1996). Alternative procedures for testing regression slope homogeneity when group error variances are unequal. *Psychological Methods, 1,* 261–277.

Diggle, P. J., Liang, K., & Zeger, S. L. (1994). *Analysis of longitudinal data.* Oxford: Clarendon Press.

Dobson, A. J. (1990). *An introduction to generalized linear models* (2nd ed.). London: Chapman and Hall.

Draper, N. R., Guttman, I., & Kanemasu, H. (1971). The distribution of certain regression statistics. *Biometrika, 58,* 295–298.

Dunlop, D. D. (1994). Regression for longitudinal data: A bridge from least squares regression. *The American Statistician, 48*(4), 299–303.

Efroymson, M. A. (1960). Multiple regression analysis. In A. Ralston & H. S. Wilf (Eds.), *Mathematical methods for digital computers* (pp. 191–203). New York: John Wiley & Sons.

Feigelson, E. D., & Babu, G. J. (Eds.). (1992). *Statistical challenges in astronomy.* New York: Springer-Verlag.

Fennessey, J., & D'Amico, R. (1980). Colinearity, ridge regession, and investigator judgment. *Social Methods and Research, 8,* 309–340.

Finkelstein, J. W., von Eye, A., & Preece, M. A. (1994). The relationship between aggressive behavior and puberty in normal adolescents: A longitudinal study. *Journal of Adolescent Health, 15,* 319–326.

Fisher, G. A. (1988). Problems in the use and interpretation of product variables. In J. S. Long (Ed.), *Common problems / proper solutions* (pp. 84–107). Newbury Park, CA: Sage.

Fisher, R. A., & Yates, R. (1963). *Statistical tables for biological, agricultural and medical research* (5th ed.). Edinburgh: Oliver-Boyd.

Fleury, P. (1991). Model II regression. *Sysnet, 8,* 2–3.

Furnival, G. M., & Wilson, R. W. (1974). Regression by leaps and bounds. *Technometrics, 16,* 499–511.

Games, P. A. (1983). Curvilinear transformations of the dependent variable. *Psychological Bulletin, 93,* 382–387.

Games, P. A. (1984). Data transformations, power, and skew: A rebuttal to Levine and Dunlap. *Psychological Bulletin, 95,* 345–347.

Goodall, C. (1983). M-estimators of location: An outline of the theory. In D. C. Hoaglin, F. Mosteller, & J. W. Tukey (Eds.), *Understanding robust and exploratory data analysis* (pp. 339–403). New York: John Wiley & Sons.

Graybill, F. A. (1976). *Theory and application of the linear model.* North Scituate, MA: Duxbury Press.

Graybill, F. A. (1983). *Matrices with applications in statistics* (2nd ed.). Belmont, CA: Wadsworth.

Gruber, M. (1989). *Regression estimators: A comparative study.* San Diego, CA: Academic Press.

Hadi, A. S., & Ling, R. F. (1998). Some cautionary notes on the use of principle components regression. *The American Statistician, 52,* 15–19.

Hartigan, J. A. (1983). *Bayes theory.* New York: Springer-Verlag.

Hettmansperger, T. P., & Sheather, S. J. (1992). A cautionary note on the method of least median squares. *The American Statistician, 46,* 79–83.

Hilgard, E. R., & Bower, G. H. (1975). *Theories of learning.* Englewood Cliffs, NJ: Prentice Hall.

Hoaglin, D. C., Mosteller, F., & Tukey, J. W. (1983). Introduction to more refined estimators. In D. C. Hoaglin, F. Mosteller, & J. W. Tukey (Eds.), *Understanding robust and exploratory data analysis* (pp. 283–296). New York: John Wiley & Sons.

Hocking, R. R. (1996). *Methods and applications of linear models.* New York: John Wiley & Sons.

Huber, P. J. (1981). *Robust statistics.* New York: John Wiley & Sons.

Huberty, C. J. (1994). A note on interpreting an R^2 value. *Journal of Educational and Behavioral Statistics, 19,* 351–356.

Isobe, T., Feigelson, E. D., Akritas, M. G., & Babu, G. J. (1990). Linear regression in astronomy i. *The Astrophysical Journal, 364,* 104–113.

Ito, P. K. (1980). Robustness of ANOVA and MANOVA test procedures. In P. R. Krishnaiah (Ed.), *Handbook of Statistics, Vol. I, Analysis of Variance* (pp. 198–236). Amsterdam: North Holland.

Jolicoeur, P. (1973). Imaginary confidence limits of the slope of the major axis of a bivariate normal distribution: A sampling experiment. *Journal of the American Statistical Association, 68,* 866–871.

Jolicoeur, P. (1991). *Introduction à la biométrie.* Montreal: Décarie.

Jolicoeur, P., & Mosiman, J. E. (1968). Intervalles de confidance pour la pente de l'axe majeur d'une distribution normal bidimensionelle. *Biométrie – Praximétrie, 9,* 121–140.

Kaskey, G., Koleman, B., Krishnaiah, P. R., & Steinberg, L. (1980). Transformations to normality. In P. R. Krishnaiah (Ed.), *Handbook of Statistics Vol. I, Analysis of Variance* (pp. 321–341). Amsterdam: North Holland.

Kermak, K. A., & Haldane, J. B. S. (1950). Organic correlation and allometry. *Biometrika, 37,* 30–41.

Kirk, R. E. (1995). *Experimental design: Procedures for the behavioral sciences* (3rd ed.). Monterey: Brooks/Cole Publishing.

Klambauer, G. (1986). *Aspects of calculus.* New York: Springer-Verlag.

Levine, D. W., & Dunlap, W. P. (1982). Power of the f test with skewed data: Should one transform or not? *Psychological Bulletin, 92,* 272–280.

Levine, D. W., & Dunlap, W. P. (1983). Data transformation, power, and skew: A rejoinder to Games. *Psychological Bulletin, 93,* 596–599.

Marazzi, A. (1980). ROBETH: A subroutine library for robust statistical procedures. In M. M. Bearritt & D. Wishart (Eds.), *Compstat 1980, proceedings in computational statistics* (pp. 577–583). Wien: Physica.

Marazzi, A. (1993). *Algorithms, routines, and S functions for robust statistics.* Pacific Grove: Wadsworth.

Mardia, K. V., Kent, J. T., & Bibby, J. M. (1979). *Multivariate analysis.* London: Academic Press.

McCullagh, P., & Nelder, J. A. (1989). *Generalized linear models* (2nd ed.). London: Chapman and Hall.

Meumann, E. (1912). *Ökonomie und Technik des Gedächtnisses.* Leipzig: Julius Klinkhardt.

Miller, A. J. (1990). *Subset selection in regression.* London: Chapman and Hall.

Morrison, D. F. (1990). *Multivariate statistical methods* (3rd ed.). New York: McGraw-Hill.

Neter, J., Kutner, M. H., Nachtsheim, C. J., & Wasserman, W. (1996). *Applied linear statistical models* (4th ed.). Chicago: Irwin.

Nichols, D. (1995a). Using categorical variables in regression. *Keywords, 56,* 10–12.

Nichols, D. (1995b). Using polytomous predictors in regression. *Keywords, 57,* 10–11.

Olkin, I., & Pratt, J. W. (1958). Unbiased estimation of certain correlation coefficients. *Annals of Mathematical Statistics, 29,* 201–211.

Pearson, K. (1901a). On lines and planes of closest fit to systems of points in space. *Philosophical Magazine, 2,* 559–572.

Pearson, K. (1901b). On the systematic fitting of curves to observations and measurements. *Biometrika, 1,* 265–303.

Pierce, M. J., & Tulley, R. B. (1988). Distances to the VIRGO and URSANAJOR clusters and a determination of h_0. *The Astrophysical Journal, 330,* 579–595.

Pope, P. T., & Webster, J. T. (1972). The use of an F-statistic in stepwise regression procedures. *Technometrics, 14*, 327–340.

Rollett, W. (1996). *Klassische, orthogonale, symmertrische und robuste symmetrische Regressionsverfahren im Vergleich.* University of Potsdam: Department of Psychology: Unpublished Master's Thesis.

Rousseeuw, P. J. (1984). Least median squares regression. *Journal of the American Statistical Association, 79*, 871–880.

Rousseeuw, P. J., & Leroy, A. M. (1987). *Robust regression and outlier detection.* New York: John Wiley & Sons.

Rovine, M. J., & von Eye, A. (1996). Correlation and categorization under a matching hypothesis. In A. von Eye & C. C. Clogg (Eds.), *Analysis of categorical variables in developmental research* (pp. 233–248). San Diego, CA: Academic Press.

Rubin, V. C., Burstein, D., & Thonnerd, N. (1980). A new relation for estimating the intrinsic luminosities of spiral galaxies. *The Astrophysical Journal, 242*, 149–152.

Ryan, T. P. (1997). *Modern regression analysis.* New York: John Wiley & Sons.

Schatzoff, M., Tsao, T., & Fienberg, S. (1968). Efficient calculation of all possible regressions. *Technometrics, 10*, 769–779.

Searle, S. R. (1971). *Linear models.* New York: John Wiley & Sons.

Searle, S. R. (1982). *Matrix algebra useful for statistics.* New York: John Wiley & Sons.

Snijders, T. A. B. (1996). What to do with the upward bias in R^2: A comment on Huberty. *Journal of Educational and Behavioral Statistics, 21*, 283–287.

Spiel, C. (1998). *Langzeiteffekte von Risiken in der frühen Kindheit [Long term effects of risks in early childhood.].* Bern: Huber.

Strömberg, G. (1940). Accidental and systematic errors in spectroscopic absolute magnitudes for dwarf go-k2 stars. *The Astrophysical Journal, 92*, 156–160.

Stuart, A., & Ord, J. K. (1991). *Kendall's advanced theory of statistics* (Vol. 2, 5th ed.). New York: Oxford University Press.

Tsutakawa, R. K., & Hewett, J. E. (1978). Comparison of two regression lines over a finite interval. *Biometrics, 34*, 391–398.

Tukey, J. W. (1977). *Exploratory data analysis.* Reading, MA: Addison-Wesley.

Venables, W. N., & Ripley, B. D. (1994). *Modern applied statistics with S-Plus.* New York: Springer-Verlag.

von Eye, A. (1983). *t* tests for single means of autocorrelated data. *Biometrical Journal, 25*, 801–805.

von Eye, A., & Rovine, M. J. (1993). Robust symmetrical autoregression in observational astronomy. In O. Lessi (Ed.), *Applications of time series analysis in astronomy and meteorology* (pp. 267–272). Padova: Università di Padova, Dipartimento di Science Statistiche.

von Eye, A., & Rovine, M. J. (1995). Symmetrical regression. *The Methodology Center Technical Report Series, 1*, The Pennsylvania State University.

von Eye, A., Sörensen, S., & Wills, S. D. (1996). Kohortenspezifisches Wissen als Ursache für Altersunterschiede im Gedächtnis für Texte und Sätze. In C. Spiel, U. Kastner-Koller, & P. Deimann (Eds.), *Motivation und Lernen aus der Perspektive lebenslanger Entwicklung* (pp. 103–119). Münster: Waxmann.

Ware, J. H. (1985). Linear models for the analysis of longitudinal studies. *The American Statistician, 39*, 95–101.

Welkowitz, J., Ewen, R. B., & Cohen, J. (1990). *Introductory statistics for the behavioral sciences* (4th ed.). San Diego, CA: Harcourt Brace Jovanovich.

Whittaker, J. (1990). *Graphical models in applied multivariate statistics.* New York: John Wiley & Sons.

Wilkinson, L., Blank, G., & Gruber, C. (1996). *Desktop data analysis with SYSTAT.* Upper Saddle River: Prentice Hall.

Wilkinson, L., Hill, M., Welna, J. P., & Birkenbeuel, G. K. (1992). *SYSTAT statistics.* Evanston, IL: Systat Inc.

Yerkes, R. M., & Dodson, J. D. (1908). The relation of strength of stimulus to rapidity of habit formation. *Journal of Comparative and Neurological Psychology, 8*, 459–482.

Yung, Y. F. (1996). Comments on Huberty's test of the squared multiple correlation coefficient. *Journal of Educational and Behavioral Statistics, 21*, 288–298.

Index